# Math Contests
# Grades 7 and 8
## (and Algebra Course 1)
### Volume 6

## School Years
## 2006-2007 through 2010-2011

### Written by

**Steven R. Conrad • Daniel Flegler • Adam Raichel**

Published by MATH LEAGUE PRESS
Printed in the United States of America

Cover art by Bob DeRosa

Phil Frank Cartoons Copyright © 1993 by CMS

First Printing, 2011

Math League Press
P.O. Box 17
Tenafly, NJ 07670-0017

ISBN 978-0-940805-19-4

# Preface

*Math Contests—Grades 7 and 8, Volume 6* is the sixth volume in our series of problem books for grades 7 and 8. The first five volumes contain the contests given in the school years 1979-1980 through 2005-2006. This volume contains the contests given from 2006-2007 through 2010-2011. (You can use the order form on page 154 to order any of our 18 books.)

This book is divided into three sections for ease of use by students and teachers. You'll find the contests in the first section. Each contest consists of 30, 35, or 40 multiple-choice questions that you can do in 30 minutes. On each 3-page contest, the questions on the 1st page are generally straightforward, those on the 2nd page are moderate in difficulty, and those on the 3rd page are more difficult. In the second section of the book, you'll find detailed solutions to all the contest questions. In the third and final section of the book are the letter answers to each contest. In this section, you'll also find rating scales you can use to rate your performance.

Many people prefer to consult the answer section rather than the solution section when first reviewing a contest. We believe that reworking a problem when you know the answer (but *not* the solution) often leads to increased understanding of problem-solving techniques.

Each school year, we sponsor an Annual 7th Grade Mathematics Contest, an Annual 8th Grade Mathematics Contest, and an Annual Algebra Course 1 Mathematics Contest. A student may participate in the contest on grade level or for any higher grade level. For example, students in grade 7 (or below) may participate in the 8th Grade Contest. *Any* student may participate in the Algebra Course 1 Contest. Starting with the 1991-92 school year, students have been permitted to use calculators on any of our contests.

**Steven R. Conrad, Daniel Flegler, & Adam Raichel, contest authors**

i

# Acknowledgments

For her continued patience and understanding, special thanks to Marina Conrad, whose only mathematical skill, an important one, is the ability to count the ways.

For demonstrating the meaning of selflessness on a daily basis, special thanks to Grace Flegler.

To Jeannine Kolbush, who did an awesome proofreading job, thanks!

# Table Of Contents

Preface . . . . . . . . . . . . . . . . . . . . . . . . . . . . . . . . . . . . . . . . . . . . . i

Acknowledgements . . . . . . . . . . . . . . . . . . . . . . . . . . . . . . . . . . . . ii

| Grade | School Year | Page for Contest | Page for Solutions | Page for Answers |
|-------|-------------|------------------|--------------------|------------------|
| 7 | 2006-07 | 5 | 73 | 138 |
| 7 | 2007-08 | 9 | 77 | 139 |
| 7 | 2008-09 | 13 | 81 | 140 |
| 7 | 2009-10 | 17 | 85 | 141 |
| 7 | 2010-11 | 21 | 89 | 142 |
| 8 | 2006-07 | 27 | 95 | 143 |
| 8 | 2007-08 | 31 | 99 | 144 |
| 8 | 2008-09 | 35 | 103 | 145 |
| 8 | 2009-10 | 39 | 107 | 146 |
| 8 | 2010-11 | 43 | 111 | 147 |
| Algebra 1 | 2006-07 | 49 | 117 | 148 |
| Algebra 1 | 2007-08 | 53 | 121 | 149 |
| Algebra 1 | 2008-09 | 57 | 125 | 150 |
| Algebra 1 | 2009-10 | 61 | 129 | 151 |
| Algebra 1 | 2010-11 | 65 | 133 | 152 |

Order Form For Contest Books (Grades 4-12) . . . . . . . . . . . . . . . . 154

# The Contests

● ● ● ● ● ● ● ● ● ● ● ● ● ● ● ● ● ● ● ● ● ●

## 2006-2007 through 2010-2011

# 7th Grade Contests

## 2006-2007 through 2010-2011

## 2006-2007 Annual 7th Grade Contest

*Tuesday, February 20 or 27, 2007*

**7**

### Instructions

- **Time** You will have only *30 minutes* working time for this contest. You might be *unable* to finish all 40 questions in the time allowed.

- **Scores** Please remember that *this is a contest, not a test*—and there is no "passing" or "failing" score. Few students score as high as 30 points (75% correct). Students with half that, 15 points, *should be commended!*

- **Format and Point Value** This is a multiple-choice contest. Each answer is an A, B, C, or D. Write each answer in the *Answers* column to the right of each question. A correct answer is worth 1 point. Unanswered questions get no credit. You **may** use a calculator.

Copyright © 2007 by Mathematics Leagues Inc.

1. By how much does 3300 exceed the sum $99+199+999+1999$?
   A) 14    B) 9    C) 4    D) 1

2. Round 0.2727 to the nearest thousandth.
   A) 0.272   B) 0.273   C) 0.2730   D) 0.2737

3. $4 \times 44$ has the same value as
   A) $16 \times 4$   B) $12 \times 33$   C) $11 \times 8$   D) $8 \times 22$

4. 80 hundredths = _?_ fifths
   A) 4    B) 8    C) 16    D) 20

5. $64 \div 8 \div 4 \times 2 =$
   A) 4    B) 8    C) 16    D) 64

6. Balloons cost 3 for $2.50 or 1 for $1.00. For $13.00, I can buy *at most*
   A) 16 balloons   B) 15 balloons
   C) 14 balloons   D) 13 balloons

7. $(5 \times 10) + (5 \times 20) + (5 \times 30) = 10 \times$ _?_
   A) $(5+50+100)$   B) $(5+20+30)$   C) $(5+10+20)$   D) $(5+10+15)$

8. The capacity of _?_ two-liter cans is half that of 16 three-liter cans.
   A) 8    B) 12    C) 16    D) 24

9. Multiplying a number by $\frac{6}{4}$ is the same as dividing it by
   A) $\frac{2}{3}$    B) $\frac{3}{2}$    C) $\frac{3}{4}$    D) $\frac{4}{3}$

10. How many positive integral factors of 60 are multiples of 6?
    A) 2    B) 3    C) 4    D) 5

11. $(3^2+4^2+5^2) \times (5^2-4^2-3^2) =$
    A) $9+16+25$   B) $6+8+10$   C) $3+4+5$   D) 0

12. If today is Tuesday, and if I first bent over backwards 563 days ago, then on which day did I first bend over backwards?
    A) Saturday    B) Friday
    C) Wednesday    D) Tuesday

13. When I subtract the product of $\frac{1}{2}$ and $\frac{1}{4}$ from their sum, I get
    A) 0    B) $\frac{3}{8}$    C) $\frac{5}{8}$    D) $\frac{3}{4}$

14. $\frac{1}{2}$ of 20 = 20% of _?_
    A) 40    B) 50    C) 80    D) 100

*Go on to the next page* ⏩ **7**

15. Which of the following has an odd number of sides?

    A) rhombus     B) octagon     C) hexagon     D) pentagon

15. D

16. Of the following, ? has the largest prime factor.

    A) 7          B) 14          C) 19          D) 20

16. D

17. What is the largest integer that could be the length of a side of an equilateral polygon whose perimeter is 100?

    A) 20          B) 25          C) 33          D) 50

17. C

18. What fraction of a week is 7 seconds?

    A) $\frac{1}{24}$     B) $\frac{1}{1440}$     C) $\frac{1}{3600}$     D) $\frac{1}{86\,400}$

18. D

19. If 1 giant apple a day keeps the doctor away, then it takes ? giant apples to keep the doctor away from March 1 through June 1.

    A) 93     B) 92     C) 91     D) 90

19. B

20. $1 \times 4 \times 9 = \sqrt{1 \times 2 \times 3 \times ?}$

    A) 1296     B) 216     C) 36     D) 6

20. A

21. The average of $\frac{5}{4}$, $\frac{7}{4}$, $\frac{9}{4}$ and $\frac{11}{4}$ equals the average of 1 and

    A) 3          B) 4          C) 9          D) 15

21. A

22. (number of dimes in \$850) = ( ? × number of quarters in \$850)

    A) 0.25     B) 1.5     C) 2.5     D) 10

22.

23. If *number* is even, then which of the following *could* be odd?

    A) $(number)^2$     B) $\sqrt{number}$     C) $2 \times number$     D) $number \div 2$

23. D

24. As shown, the square at the right is divided into 8 identical isosceles triangles. How many different right angles does the figure contain?

    A) 20          B) 16          C) 12          D) 8

24. A

25. $2^3 \times 3^3 \times 5^3 = 3 \times 4 \times 5 \times 6 \times ?$

    A) 100     B) 75     C) 50     D) 25

25. B

26. I've dieted for 30 days. That's 40% of the ? days I'll stay on my diet.

    A) 12     B) 42     C) 50     D) 75

26. D

27. $\frac{2}{3} \times \frac{3}{2} \times \frac{2}{3} \times \frac{3}{2} \times \frac{2}{3} \times \frac{3}{2} \times \frac{2}{3} =$

    A) $\frac{1}{3}$     B) $\frac{2}{3}$     C) 1     D) $\frac{3}{2}$

27. B

28. $5^4$ is ? % of $10^4$.

    A) 0.5          B) 6.25          C) 25          D) 50

28. B

29. What is the area of a semicircular region whose diameter is 16?

    A) $32\pi$    B) $64\pi$    C) $128\pi$    D) $256\pi$

    29. A

30. If $r(N)$ means the reciprocal of $N$, then $r(r(2) \div r(3) \div r(4)) =$

    A) $\frac{2}{3}$    B) $\frac{3}{8}$    C) $\frac{1}{6}$    D) $\frac{1}{24}$

    30. C

31. Cupid shot his arrow 10 000 times. His success rate was $\sqrt{1\%}$. How many of Cupid's shots were a success?

    A) 1    B) 10    C) 100    D) 1000

    31. B

32. The largest factor of 10 000 is

    A) 5    B) 1000    C) 5000    D) 10 000

    32. D

33. Today, 2 horses each chased 4 cows, each of which chased 8 pigs, each of which chased 16 chickens. No animal was chased by 2 animals. Altogether, how many animals took part in this chase?

    A) 30    B) 512    C) 1024    D) 1098

    33. D

34. If the measure of each interior angle of 9-sided polygon $P$ is less than 180, then $P$ has at most _?_ different *pairs* of parallel sides.

    A) 2    B) 3    C) 4    D) 8

    34. C

35. Which *cannot* be the smallest of 5 consecutive integers whose product is divisible by 2, 5, and 9?

    A) 5    B) 4    C) 3    D) 2

    35. D

36. Cory averaged 85 on fifteen tests. He got two 80s, five 85s, and three 90s. What is the average of Cory's other five tests?

    A) 82    B) 84    C) 85    D) 86

    36. B

37. $\frac{2007}{2006} \div \underline{\ ?\ } = \frac{2006}{2007}$

    A) $\frac{2007^2}{2006^2}$    B) $\frac{2006^2}{2007^2}$    C) $\frac{2 \times 2006^2}{2007^2}$    D) 1

    37. A

38. The least common denominator of $\frac{1}{2}, \frac{2}{3}, \frac{3}{4}, \frac{4}{5}, \frac{5}{6}, \frac{6}{7}, \frac{7}{8}, \frac{8}{9}$, and $\frac{9}{10}$ is

    A) 10    B) 1260    C) 2520    D) 3 628 800

    38. D

39. Of the first 999 positive integers, _?_ are primes or multiples of primes.

    A) 1    B) 499    C) 998    D) 999

    39. C

40. In the sum shown at the right, if $R$, $S$, $T$, and $U$ represent digits, then $R+S+T+U$ must equal

    A) $2 \times (R+S)$    B) $2 \times (S+U)$    C) $2 \times (T+U)$    D) $2 \times T$

    $$\begin{array}{r} RSTU \\ + UTSR \\ \hline TTTT \end{array}$$

    40. D

*The end of the contest* ✍ **7**

# 2007-2008 Annual 7th Grade Contest

*Tuesday, February 19 or 26, 2008*

**7**

## Instructions

- **Time** You will have only *30 minutes* working time for this contest. You might be *unable* to finish all 40 questions in the time allowed.

- **Scores** Please remember that *this is a contest, not a test*—and there is no "passing" or "failing" score. Few students score as high as 30 points (75% correct). Students with half that, 15 points, *should be commended!*

- **Format and Point Value** This is a multiple-choice contest. Each answer is an A, B, C, or D. Write each answer in the *Answers* column to the right of each question. A correct answer is worth 1 point. Unanswered questions get no credit. You **may** use a calculator.

Copyright © 2008 by Mathematics Leagues Inc.

1. $10 \times 2008 + 2008 =$
   A) 4026　　B) $10 \times 2009$　C) $10 \times 4016$　D) $11 \times 2008$

   **1.** D

2. At the annual car wash, a customer left 3 coins as a tip. The total value of this tip *cannot* be _?_ ¢.
   A) 30　B) 40　C) 50　D) 60

   **2.** C

3. The least common multiple of the numbers 2, 4, and 8 is
   A) 2　B) 8　C) 14　D) 64

   **3.** B

4. 10 hundredths + 10 hundredths =
   A) 0.200　　B) 0.020　　C) 0.110　　D) 0.100

   **4.** A

5. The product of two integers is 35. If each > 1, their sum is
   A) 8　　B) 12　　C) 18　　D) 36

   **5.** B

6. Of the following, which has the greatest value?
   A) $0.1 \times 1.1 \times 1$　B) $0.1 + 1.1 \times 1$　C) $1.1 + 0.1 \times 1$　D) $0.1 + 1.1 + 1$

   **6.** D

7. I can cut at most _?_ pieces $1\frac{1}{4}$ cm long from a string 20 cm long.
   A) 15　　B) 16　　C) 18　　D) 25

   **7.** B

8. What is the sum of my three test grades, if their average is 90?
   A) 270　　B) 180　　C) 90　　D) 30

   **8.** A

9. We held our Splash Down Party at a pool in the shape of a square with area 36 m². What is the perimeter of this pool?
   A) 18 m　B) 24 m　C) 36 m　D) 81 m

   **9.** B

10. Al is 5 km from Bob. Dan is 4 km from Bob. The distance between Al and Dan is *at least*
    A) 1 km　B) 2 km　C) 3 km　D) 4 km

    **10.** A

11. The ratio of hours in a day to days in an hour is
    A) 1:24　　B) 24:1　C) 48:1　　D) 576:1

    **11.** D

12. Of the following, the one with the largest prime factor is
    A) 49　　B) 51　　C) 58　　D) 65

    **12.** C

13. $\frac{1}{2}$ of $\frac{1}{3} = \frac{1}{3}$ of _?_　　　A) $\frac{1}{6}$ B) $\frac{1}{4}$ C) $\frac{1}{3}$ D) $\frac{1}{2}$

    **13.** D

14. The product of _?_ and its square is 64.
    A) 2　　B) 4　　C) 6　　D) 8

    **14.** B

15. How many integers are between $\frac{10}{9}$ and $\frac{100}{9}$?
    A) 8　　B) 9　　C) 10　　D) 90

    **15.** D

*Go on to the next page* �\|▶ **7**

16. $\sqrt{9} + \sqrt{16} = \sqrt{25} + \underline{\ ?\ }$

    A) $\sqrt{0}$      B) $\sqrt{2}$      C) $\sqrt{4}$      D) $\sqrt{49}$

16.

17. How many of the numbers 0.1, 0.9, 1.0, 1.9 exceed their reciprocals?

    A) 1      B) 2      C) 3      D) 4

17.

18. If my canned food drive ended at 4:18 P.M. and was half over at 3:26 P.M., then it began at $\underline{\ ?\ }$ P.M.

    A) 2:18      B) 2:32

    C) 2:34      D) 2:44

18.

19. The numbers $\underline{\ ?\ }$ *cannot* be lengths of a triangle's sides.

    A) 4, 5, 6      B) 3, 4, 5

    C) 2, 3, 4      D) 1, 2, 3

19.

20. $\dfrac{31}{16} + \dfrac{32}{16} + \dfrac{33}{16} = \dfrac{?}{8}$      A) 48   B) 36   C) 16   D) 12

20.

21. The number $\dfrac{36}{54}$ is *not* reducible to a fraction whose denominator is

    A) 6      B) 8      C) 12      D) 15

21.

22. The sum of 2 consecutive whole numbers *cannot* be

    A) 1      B) prime      C) odd      D) even

22.

23. Which number below doubles when I increase its numerator by 12?

    A) $\dfrac{3}{41}$      B) $\dfrac{6}{41}$      C) $\dfrac{12}{41}$      D) $\dfrac{24}{41}$

23.

24. The measures of 2 angles in an isosceles triangle could be

    A) 85°, 50°      B) 80°, 55°      C) 75°, 35°      D) 70°, 40°

24.

25. 10% of 10% = 100% of $\underline{\ ?\ }$

    A) 1      B) $\dfrac{1}{10}$      C) $\dfrac{1}{100}$      D) $\dfrac{1}{1000}$

25.

26. When fully expanded, $10^{2008}$ has $\underline{\ ?\ }$ digits.

    A) 2009    B) 2008    C) 2007    D) 20 080

26.

27. Ann is 2 cm taller than Bob, who is 2 cm taller than Carl. If the average height of all three is 150 cm, then Carl is $\underline{\ ?\ }$ cm tall.

    A) 152      B) 148      C) 147      D) 146

27.

28. $\dfrac{1+2+3+4+5}{2+4+6+8+10} = \dfrac{6+7}{?}$

    A) 12      B) 14      C) 26      D) 27

28.

29. (least perfect square > 2008) − (greatest perfect square < 2008) =

    A) 88      B) 89      C) 90      D) 91

29.

*Go on to the next page* ➠ **7**

30. A circle *cannot* pass through exactly __?__ of a square's vertices.

    A) 4          B) 3          C) 2          D) 1

    30.

31. Of the inequalities below, only __?__ is correct.

    A) $\frac{7}{8} < \frac{8}{17} < \frac{6}{11}$  B) $\frac{6}{11} < \frac{8}{17} < \frac{7}{8}$  C) $\frac{6}{11} < \frac{7}{8} < \frac{8}{17}$  D) $\frac{8}{17} < \frac{6}{11} < \frac{7}{8}$

    31.

32. The volume of a cube with a surface area of 9600 is __?__ times the volume of a cube with a surface area of 96.

    A) 10     B) 100     C) 1000     D) 10 000

    32.

33. To buy a $45 gift, I paid 25% right away, and I paid the rest in 5 equal payments of

    A) $6.75  B) $7.25  C) $9.00  D) $11.25

    33.

34. If $N$ is a prime, which of the following is *never* a prime?

    A) $N+3$          B) $N+5$
    C) $N+7$          D) $N+9$

    34.

35. The value of $2 \times \frac{1}{3} \times 4 \times \frac{1}{5} \times 6 \times \frac{1}{7} \times \ldots \times 48 \times \frac{1}{49}$ is

    A) $< 1$     B) $> 1$     C) $= 1$     D) $= 0$

    35.

36. Of 400 kids at my school, 30% have a cat and 75% have a dog. Of the kids with a cat, 65 also have a dog. How many of the 400 kids have neither a cat nor a dog?

    A) 45     B) 98     C) 129     D) 335

    36.

37. $8^8 \times 4^4 \times 2^2 =$

    A) $2^{64}$     B) $2^{34}$     C) $2^{26}$     D) $2^{14}$

    37.

38. $\sqrt{\sqrt{\sqrt{100 \times 100 \times 100 \times 100}}} =$

    A) 1          B) $\sqrt{10}$          C) 10          D) 100

    38.

39. The sides of a regular octagon are numbered clockwise from 1 through 8. After the octagon is rotated 1575° clockwise, which side is in the position that side 3 occupied before the rotation?

    A) side 2    B) side 4    C) side 6    D) side 8

    39.

40. If I swam 1 more lap each day than I swam the day before, and I averaged 58 laps daily last May, then I swam __?__ laps last May 1.

    A) 42          B) 43          C) 44          D) 45

    40.

*The end of the contest* ✍ **7**

**Visit our Web site at http://www.mathleague.com**

Solutions on Page 77 • Answers on Page 139

# 2008-09 Annual 7th Grade Contest

*Tuesday, February 17 or 24, 2009*

**7**

## Instructions

- **Time** You will have only *30 minutes* working time for this contest. You might be *unable* to finish all 40 questions in the time allowed.

- **Scores** Please remember that *this is a contest, not a test*—and there is no "passing" or "failing" score. Few students score as high as 30 points (75% correct). Students with half that, 15 points, *should be commended!*

- **Format and Point Value** This is a multiple-choice contest. Each answer is an A, B, C, or D. Write each answer in the *Answers* column to the right of each question. A correct answer is worth 1 point. Unanswered questions get no credit. You **may** use a calculator.

13

1. $96 + 4 + 97 + 3 + 98 + 2 + 99 + 1 = 4 \times$ ?
   A) 50     B) 95     C) 96     D) 100

   1.

2. I can make 6 equal stacks from 200 dimes if ? dimes are left over.
   A) 1     B) 2     C) 3     D) 4

   2.

3. The hour hand of a standard 12-hour circular clock moves a total of 60° in ? hours.
   A) 1    B) 2    C) 3    D) 6

   3.

4. The average number of days per year in the 4 years from 2009 through 2012 is
   A) 365.00     B) 365.10
   C) 365.25     D) 365.75

   4.

5. What is $\frac{1}{3}$ of $\frac{1}{4}$ of 300?
   A) 25    B) 75    C) $\frac{1}{400}$    D) $\frac{1}{3600}$

   5.

6. By how many halves does 4.5 exceed 1.5?
   A) 1     B) 3     C) 6     D) 9

   6.

7. Of the following, which is a multiple of both 4 and 6?
   A) 1246     B) 2412     C) 4664     D) 6424

   7.

8. $1 \times 10 \times 100 \times 1000 =$
   A) 1111     B) 11 110     C) 1 000 000     D) 1 111 000 000

   8.

9. $4 \times (4+5) + 5 \times (4+5) =$
   A) $9^2$     B) 99     C) $9 + 9^2$     D) $9^3$

   9.

10. 15 is the sum of *at most* ? consecutive positive integers.
    A) 2    B) 3    C) 4    D) 5

    10.

11. Chef Smile is smiling because his hourly salary was raised 25%, from \$40 to
    A) \$15    B) \$30    C) \$50    D) \$65

    11.

12. Divide a prime by an even number. The result of this division *cannot* be
    A) even     B) odd
    C) whole     D) 1

    12.

13. $18^2 \times 9 = 9^2 \times$ ?
    A) 18    B) $2^2$    C) $3^2$    D) $6^2$

    13.

14. What fractional part of eight is eight-hundredths?
    A) $\frac{1}{100}$     B) $\frac{8}{100}$     C) $\frac{64}{100}$     D) $\frac{8}{10}$

    14.

*Go on to the next page* ⟫ **7**

15. As shown, $\triangle ABC$ and $\triangle ACD$ are both equilateral and have side $\overline{AC}$ in common. What is $m\angle BCD$?
    A) 60    B) 120    C) 150    D) 180

16. We walked the same path up a hill, then down, with no break. We began at 3:15 P.M. and finished at 4:30 P.M. We walked the downhill part twice as fast as the uphill part. We began to walk downhill at
    A) 3:40 P.M.    B) 3:55 P.M.
    C) 4:00 P.M.    D) 4:05 P.M.

17. 1 km ÷ 1 cm =
    A) 100 000    B) 10 000
    C) 1000    D) 100

18. 3 thousand + 30 thousandths =
    A) 33 000    B) 3000.03    C) 3000.003    D) 0.330

19. What is the product of all the whole-number factors of 24?
    A) 24    B) $24^2$    C) $24^3$    D) $24^4$

20. $\sqrt{16} \times \sqrt{8} \times \sqrt{4} \times \sqrt{2} = 16 \times$ ?
    A) 2    B) 4    C) 8    D) 64

21. $12 + 72 \div 6 \times 2 =$
    A) 7    B) 18    C) 28    D) 36

22. $\frac{11}{10} + \frac{101}{100} + \frac{1001}{1000} =$
    A) $3\frac{1}{10}$    B) $3\frac{3}{10}$    C) $3\frac{1}{1000}$    D) $3\frac{111}{1000}$

23. $24 \times 26 \times 28 \times 30 = 12 \times 13 \times 14 \times 15 \times$ ?
    A) 16    B) 12    C) 4    D) 2

24. What percent of 60 is 20% of 30?
    A) 1%    B) 10%    C) 40%    D) 100%

25. How many more thirds than halves equal 30?
    A) 5    B) 15    C) 20    D) 30

26. The distance between 2 points on a circle with area $\pi$ is *at most*
    A) 0.5    B) 1    C) 2    D) 4

27. The reciprocal of $\left(\frac{1}{2}+2\right)$ is
    A) $\frac{2}{5}$    B) $\frac{2}{3}$    C) $\frac{3}{2}$    D) $\frac{5}{2}$

28. If 2 *different* polygons average 5 sides each, one polygon could be
    A) a pentagon    B) a hexagon    C) an octagon    D) a decagon

29. Just as $12 = 2 \times 2 \times 3$ is the product of 3 primes, not all different, $1\,000\,000\,000\,000$ is the product of _?_ primes, not all different.
    A) 20      B) 24      C) 48      D) 50 | 29.

30. I rode 24 km in 3 hrs. Al went twice as far in half the time when he rode at the rate of
    A) 8 km/hr      B) 16 km/hr
    C) 32 km/hr      D) 64 km/hr | 30.

31. Which can be written as the sum of 3 consecutive odd numbers?
    A) 100      B) 99
    C) 98      D) 97 | 31.

32. 20% of 0.5% =
    A) 0.1%    B) 1%    C) 10%    D) 10 | 32.

33. Dividing $\frac{1}{2}$ by $\frac{2}{3}$ yields the same result as multiplying $\frac{2}{3}$ by
    A) 2      B) $\frac{9}{8}$      C) $\frac{3}{4}$      D) $\frac{1}{2}$ | 33.

34. Exactly _?_ different even whole numbers are factors of $2 \times 3 \times 5 \times 7$.
    A) 1      B) 6      C) 7      D) 8 | 34.

35. You earn 0.5% of what I earn. I earn _?_ % of what you earn.
    A) 2      B) 200      C) 2000      D) 20 000 | 35.

36. $\left(1-\frac{1}{2}\right) \times \left(1-\frac{1}{3}\right) \times \left(1-\frac{1}{4}\right) \times \ldots \times \left(1-\frac{1}{100}\right) =$
    A) $\frac{1}{100}$      B) $\frac{1}{99}$      C) $\frac{9}{10}$      D) $\frac{99}{100}$ | 36.

37. Using whole number ages, if Al : Ed = 3 : 5, then Al + Ed *could* be
    A) 62      B) 72      C) 82      D) 92 | 37.

38. If a square has whole-number side-lengths, its area could be
    A) $7^3$    B) $8^3$    C) $9^3$    D) $10^3$ | 38.

39. Of 50 people, if 20 sing, 20 dance, and 15 do both, then _?_ do neither.
    A) 5    B) 10    C) 15    D) 25 | 39.

40. Grandma and Grandpa danced for 20 minutes one day; for each of the next 91 days, they danced for one minute longer than they had on the day before. On those 92 days, they danced for a total of _?_ minutes.
    A) 1820    B) 5106    C) 6026    D) 6072 | 40.

*The end of the contest* ✍ **7**

**SEVENTH GRADE MATHEMATICS CONTEST**

Math League Press, P.O. Box 17, Tenafly, New Jersey 07670-0017

# 2009-2010 Annual 7th Grade Contest

*Tuesday, February 16 or 23, 2010*

**7**

## Instructions

- **Time** Do *not* open this booklet until you are told by your teacher to begin. You might be *unable* to finish all 40 questions in the 30 minutes allowed.

- **Scores** Please remember that *this is a contest, and not a test*—there is no "passing" or "failing" score. Few students score as high as 30 points (75% correct). Students with half that, 15 points, *should be commended!*

- **Format, Point Value, & Eligibility** Every answer is an A, B, C, or D. Write answers in the *Answers* column. A correct answer is worth 1 point. Unanswered questions get no credit. You **may** use a calculator.

1. A large square is divided into 25 small squares as shown. What percent of the large square is shaded?

   A) 5%  B) 20%  C) 25%  D) 50%

   *1.* B

2. When __?__ is divided by 4, the quotient is 18 and the remainder is 2.

   A) 26    B) 56    C) 70    D) 74

   *2.* D

3. To the nearest tenth, 2.345 is

   A) 2.3  B) 2.34  C) 2.35  D) 2.5

   *3.* A

4. The intersection of $\{a, b, c\}$ and $\{b, c, d\}$ is

   A) { }  B) $\{b, c\}$  C) $\{a, d\}$  D) $\{a, b, c, d\}$

   *4.* B

5. How many two-digit integers are greater than 20?

   A) 81  B) 80  C) 79  D) 78

   *5.* C

6. $(8 \times 6 + 2) \div 2 =$

   A) $4 \times 3 + 1$  B) $4 \times 6 + 1$  C) $4 \times 6 + 2$  D) $8 \times 6 + 1$

   *6.* B

7. Which of the following is a prime number?

   A) 81  B) 83  C) 87  D) 99

   *7.* B

8. 1% of $2010 is

   A) $0.201  B) $2.01  C) $20.10  D) $201.00

   *8.* C

9. $\sqrt{36} - \sqrt{25} =$

   A) $\sqrt{1}$  B) $\sqrt{11}$  C) $\sqrt{14}$  D) $\sqrt{16}$

   *9.* A

10. Eight sisters are given a roll of quarters to share equally. There are 40 quarters in a roll. After three of the sisters take their shares of the quarters, what will be the total value of the quarters remaining in the roll?

    A) $0.15  B) $1.25  C) $3.75  D) $6.25

    *10.* D

11. $2^4 + 2^4 =$

    A) $2^5$  B) $4^8$  C) $4^4$  D) $2^8$

    *11.* A

12. The ratio of the number of sides of a hexagon to the number of sides of an octagon is

    A) 7:8  B) 5:6  C) 3:4  D) 1:2

    *12.* C

13. The average of $\frac{1}{2}$ and $\frac{1}{3}$ is

    A) $\frac{5}{6}$  B) $\frac{2}{5}$  C) $\frac{1}{6}$  D) $\frac{5}{12}$

    *13.* D

14. $(6 \times 12 \times 18 \times 24) \div (3 \times 6 \times 9 \times 12) =$

    A) 16  B) 8  C) 2  D) 0

    *14.* A

15. If a circle's radius is divided by its circumference, the quotient is

    A) $2\pi$  B) $\pi$  C) $\frac{1}{\pi}$  D) $\frac{1}{2\pi}$

    *15.* D

16. Of the following, which most nearly equals 23.4 divided by 0.5?

    A) 12     B) 24     C) 36     D) 48

16. D

17. If a rectangle has an area of 40 m² and a perimeter of 28 m, then the length of the longest side of the rectangle is

    A) 12 m    B) 10 m    C) 8 m    D) 5 m

17. B

18. 300% of 30 is

    A) 0.9     B) 9     C) 90     D) 9000

18. C

19. Of the following, which is the smallest number?

    A) $\frac{1}{3}$     B) $\frac{210}{60}$     C) $\frac{990}{300}$     D) $\frac{1001}{3000}$

19. C

20. On a map of the Andes, 2 cm represent 6000 km. How many cm represent 300 km?

    A) 0.1     B) 0.5     C) 10     D) 20

20. A

21. One vertex of an equilateral triangle is on a line, as shown. What is $m\angle A + m\angle B$?

    A) 60     B) 90     C) 120     D) 180

21. C

22. The alphabet is written in order over and over until 300 letters are written. What is the 300th letter written?

    A) K          B) L          C) M          D) N

    260
    286

22. D

23. After the first 4 tests of the year, Lana had an average score of 75. After the first 5, her average was 80. What was her score on the 5th test?

    A) 100     B) 95     C) 85     D) 80

23. A

24. 10 m + 10 cm =

    A) 11 m    B) 10.1 m    C) 10.01 m    D) 10.001 m

    TIC...
    TIC...
    TIC...
    TIC..

24. C

25. 18:12 =

    A) 2:3     B) 16:10     C) 12:18     D) 36:24

25. D

26. $45^3 =$

    A) $3^3 \times 5^3$     B) $4^3 \times 5^3$     C) $3^6 \times 5^3$     D) $3^8 \times 5^3$

26. C

27. What is the correct time 56 minutes after 10:56 AM?

    A) 10:00 AM     B) 11:52 AM     C) 11:56 AM     D) 12:02 PM

27. B

28. The square of the reciprocal of my favorite whole number is $\frac{1}{9}$. What is the cube of my favorite whole number?

    A) $\frac{1}{27}$     B) $\frac{1}{9}$     C) 9     D) 27

28. D

29. A train that travels at 240 km/hr. will travel ___?___ m in 30 seconds.

    A) 2          B) 8          C) 2000          D) 8000

29. C

Go on to the next page ))) ➡ 7

**Answers**

30. On a flat surface, the distance from point $A$ to point $B$ is 4 cm, and the distance from point $B$ to point $C$ is 7 cm. Which of the following *cannot* be the distance from point $A$ to point $C$? — 30.

A) 2 cm  B) 3 cm  C) 6 cm  D) 10 cm

31. In the correct multiplication shown at right, R, S, and T are different non-zero digits, and T > S > R. Which is the value of S? — 31.

$$\begin{array}{r} \text{RS} \\ \underline{\times\ \text{S}} \\ \text{1TS} \end{array}$$

A) 1  B) 4  C) 5  D) 6

32. If the shortest side of an isosceles triangle is 10, and the difference in lengths between two sides is also 10, then the triangle's perimeter is — 32.

A) 60  B) 50  C) 40  D) 30

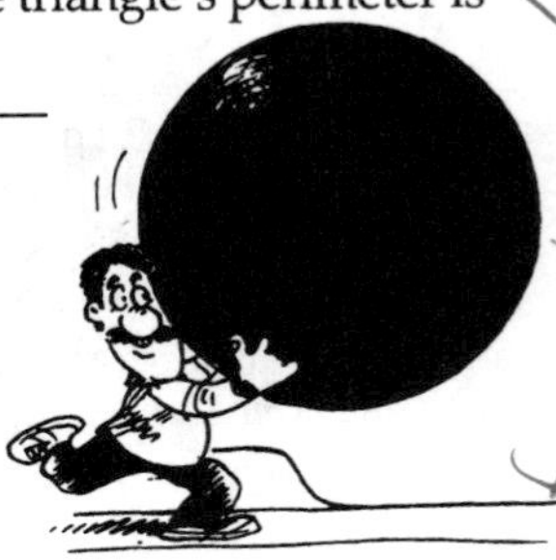

33. $\frac{9}{2}$ is how much more than $\frac{2}{9}$? — 33.

A) $\frac{18}{77}$  B) $\frac{14}{18}$  C) $\frac{18}{14}$  D) $\frac{77}{18}$

34. How many positive integers less than 200 can be written as the sum of two positive even integers? — 34.

A) 101  B) 100  C) 99  D) 98

35. A wheel of radius 2 m rolls a distance of $200\pi$ m in 2 minutes. At this rate, how many full revolutions will the wheel make in one hour? — 35.

A) 100  B) 200  C) 1500  D) 3000

36. The product of the first 2010 prime numbers is divisible by — 36.

A) 210  B) 260  C) 420  D) 520

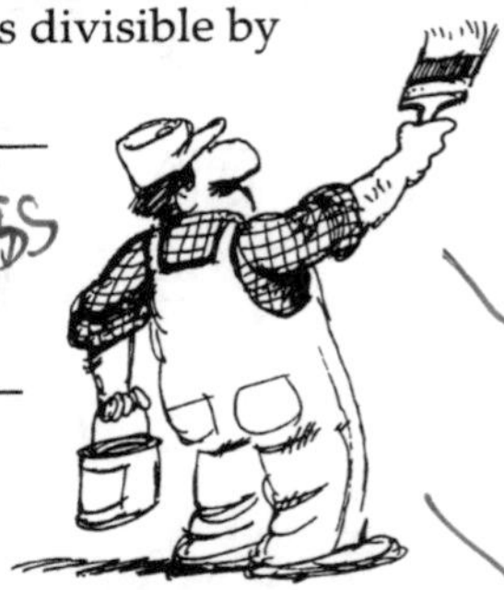

37. Of 180 paintings, 110 have blue borders and 90 have red borders. If 25 paintings have neither color border, how many have both colors? — 37.

A) 25  B) 45  C) 55  D) 65

38. The difference between the sum of the 50 smallest positive even integers and the sum of the 50 smallest positive odd integers is — 38.

A) 100  B) 50  C) 25  D) 1

39. Of the following, which has a ones digit of 2? — 39.

A) $2^{2009}$  B) $2^{2010}$  C) $2^{2011}$  D) $2^{2012}$

40. Gwen spent $\frac{1}{7}$ of her money on food, and $\frac{1}{3}$ of her remaining money on clothes. She then had \$36 left. How much did she spend on food? — 40.

A) \$24  B) \$18  C) \$12  D) \$9

*The end of the contest* 17

**Visit our Web site at http://www.mathleague.com**

Solutions on Page 85 • Answers on Page 141

# 2010-2011 Annual 7th Grade Contest

*Tuesday, February 15 or 22, 2011*

**7**

## Instructions

- **Time** Do *not* open this booklet until you are told by your teacher to begin. You might be *unable* to finish all 35 questions in the 30 minutes allowed.

- **Scores** Please remember that *this is a contest, and not a test*—there is no "passing" or "failing" score. Few students score as high as 28 points (80% correct). Students with half that, 14 points, *should be commended!*

- **Format, Point Value, & Eligibility** Every answer is an A, B, C, or D. Write answers in the *Answers* column. A correct answer is worth 1 point. Unanswered questions get no credit. You **may** use a calculator.

1. $\dfrac{1}{2011} \times 2011^2 =$

   A) 2013     B) 2011     C) 2     D) 1

   1. B

2. $(4+3) \times (5+2) \times (6+1) =$

   A) $3 \times 7$   B) $7 \times 7$   C) $3^7$   D) $7^3$

   2. D

3. Ben finished wrapping 30 boxes at 1:30 PM. If Ben wrapped 1 box every 5 minutes, then he started wrapping the boxes at __?__ AM.

   A) 10:30   B) 11:00   C) 11:30   D) 11:50

   3. B

4. $2\dfrac{3}{4} + 3\dfrac{4}{5} =$

   A) $5\dfrac{3}{5}$    B) $5\dfrac{11}{20}$    C) $6\dfrac{3}{5}$    D) $6\dfrac{11}{20}$

   4. D

5. The ones digit of the cube of 432 is

   A) 8     B) 6     C) 4     D) 2

   5. A

6. If each of the following numbers is rounded to the nearest whole number and then divided by 3, which has the greatest remainder?

   A) 14.45    B) 15.82    C) 16.39    D) 17.99

   6. A

7. One prime factor of 351 is 3. Another prime factor of 351 is

   A) 7     B) 13     C) 39     D) 117

   7. B

8. The measures of the smallest and largest angles of a right triangle could differ by

   A) 1°    B) 30°    C) 62°    D) 91°

   8.

9. Mr. L. C. M. refills his drink every 45 minutes and his pool every 105 minutes. If he refills both at 1:00 PM, then he next refills both at the same time at

   A) 3:30 PM    B) 4:30 PM    C) 6:15 PM    D) 7:15 PM

   9. C

10. If 6 cronks = 14 crunks, then 9 cronks = __?__ crunks.

    A) 24     B) 21     C) 20     D) 17

    10. B

11. When 20 is divided by 40%, the quotient is

    A) $\dfrac{1}{2}$     B) 8     C) 25     D) 50

    11. B

12. What is the area of the shaded region of the square?

    A) 8   B) 16   C) 28   D) 36

    12. C

13. How many prime numbers are between 80 and 90?

    A) 1    B) 2    C) 3    D) 4

    13. B

| | **Answers** |
|---|---|

14. Of 20 chess pieces on a board, 12 are white and the others are black. The ratio of black pieces to all pieces is

A) 1:2  B) 1:4  C) 2:3  D) 2:5

14.

15. Add the cube and the square of _?_ to get 12.

A) 2  B) 4  C) 16  D) 64

15.

16. My seven Larry Rotter books have 300 pages each, and my five Chronicles of Blarnia books have 324 pages each. What is the average number of pages per book in all 12 books?

A) 310  B) 312  C) 314  D) 316

16.

17. What is the product of the reciprocals of the first 3 integers that are squares of positive integers?

A) $\frac{1}{576}$  B) $\frac{1}{36}$  C) 36  D) 576

17.

18. A rope 12 m long is cut into 4 pieces of equal length. Each piece is wrapped exactly once around the circumference of one of four identical car wheels. The diameter of one wheel is _?_ m.

A) 6π  B) 3π  C) $\frac{3}{\pi}$  D) $\frac{3}{2\pi}$

18.

19. The sum of four consecutive whole numbers is 110. What is the sum of the least and the greatest of the four numbers?

A) 53  B) 55  C) 57  D) 58

19.

20. If one angle of a triangle is acute, and a second angle of the triangle is obtuse, then the third angle of the triangle must be

A) acute  B) obtuse  C) right  D) scalene

20.

21. A square and a rectangle share a side of length 4 as shown. The area of the entire figure is 64. The perimeter of the entire figure is

A) 52  B) 48  C) 44  D) 40

21.

22. What time is it exactly 1 440 000 minutes after 10 AM?

A) 10 AM  B) 11 AM  C) 10 PM  D) 11 PM

22.

23. The ratio 6:8.4 is equivalent to

A) 2:4.4  B) 3:5  C) 5:7  D) 6.8:4

23.

24. One angle of a parallelogram is 5 times another angle of the parallelogram. The measure of the largest angle of the parallelogram is

A) 100°  B) 120°  C) 150°  D) 160°

24.

25. Joy walks 60 m to a slide in 180 sec. and slides down 9 times faster than she walks. If it takes 30 sec. to slide down, how long is the slide?

    A) 90 m    B) 180 m   C) 270 m   D) 810 m

26. If $\frac{2}{3}$ of a number is $\frac{1}{2}$, then $\frac{1}{6}$ of the number is

    A) $\frac{1}{8}$     B) $\frac{2}{9}$     C) $\frac{3}{16}$     D) $\frac{4}{15}$

27. What percent of 24 seconds is 4 hours?

    A) 600%       B) 3600%       C) 60 000%       D) 360 000%

28. Abe's height is 40% greater than Bo's height, and Bo's height is 25% less than Cal's height. Abe's height is what percent of Cal's height?

    A) 25%       B) 85%       C) 95%       D) 105%

29. The least integer power of 12 that is divisible by $18^{180}$ is

    A) $12^{120}$       B) $12^{180}$       C) $12^{240}$       D) $12^{360}$

30. Subtract the sum of all even integers between 19 and 121 from the sum of all odd integers between 20 and 122. What is the difference?

    A) 1       B) 51       C) 100       D) 101

31. What is the volume of a rectangular box if three of its faces have areas of 30, 70, and 84?

    A) 184    B) 368    C) 420    D) 176 400

32. The sum of 4 consecutive even integers *cannot* be

    A) 4     B) 12     C) 16     D) 20

33. If the product of all prime numbers between 1 and 210 is divided by 210, the remainder is

    A) 0     B) 3     C) 7     D) 21

34. Amy picks 3 of the 7 colors of the rainbow, but she doesn't pick red with green, and she doesn't use blue at all. How many different combinations of 3 colors can Amy pick?

    A) 10       B) 16       C) 20       D) 24

35. The length of one side of a triangle is between 9 and 11. The perimeter of the triangle could be

    A) 11       B) 16       C) 18       D) 38

*The end of the contest*   7

# 8th Grade Contests

## 2006-2007 through 2010-2011

# 2006-2007 Annual 8th Grade Contest

*Tuesday, February 20 or 27, 2007*

## Instructions

- **Time** You will have only *30 minutes* working time for this contest. You might be *unable* to finish all 40 questions in the time allowed.

- **Scores** Please remember that *this is a contest, not a test*—and there is no "passing" or "failing" score. Few students score as high as 30 points (75% correct). Students with half that, 15 points, *should be commended!*

- **Format and Point Value** This is a multiple-choice contest. Each answer is an A, B, C, or D. Write each answer in the *Answers* column to the right of each question. A correct answer is worth 1 point. Unanswered questions get no credit. You **may** use a calculator.

Answers

1. Of the following, which has the least value?

A) $\frac{2}{3}$    B) $\frac{3}{5}$    C) $\frac{4}{7}$    D) $\frac{5}{9}$

1. D

2. I bought 27 fifty-cent stamps and got _?_ change from a $20 bill.

A) $18.65    B) $13.50    C) $10.35    D) $6.50

2. D

3. $\frac{4}{5} \times \frac{3}{4} \times \frac{2}{3} \times \frac{1}{2} \times \frac{0}{1} =$

A) 1    B) $\frac{1}{5}$    C) 0    D) $-\frac{1}{5}$

3. C

4. I first wore my headphones on July 1, a Friday. I last wore them on the last day of July, a

A) Saturday    B) Sunday
C) Monday    D) Tuesday

4. B

5. $(10 \times 0.1) \times (100 \times 0.01) \times (1000 \times 0.001) =$

A) 1    B) 0.1    C) 0.01    D) 0.001

5. A

6. The sum of the 99 smallest positive integers and the 99 largest negative integers is

A) 0    B) 4950    C) 9900    D) 10 000

6. A

7. I saved 2 dimes for every 3 nickels. I saved _?_ dimes and 90 nickels.

A) 15    B) 30    C) 60    D) 90

7. C

8. $\frac{1}{2} + \frac{3}{4} = \frac{5}{?}$

A) 4    B) 6    C) 7    D) 12

8. A

9. If $\angle O$ is obtuse and $\angle A$ is acute, $m\angle O - m\angle A$ can *never* equal

A) 90°    B) 89°    C) 1°    D) 0°

9. D

10. I buried my head in the ground 120 times in 15 minutes. That's an average of _?_ times every 3 minutes.

A) 8    B) 12    C) 24    D) 30

10. C

11. The reciprocal of 10 exceeds the reciprocal of 100 by

A) 0.9    B) 0.09    C) 0.10    D) 10

11. B

12. $(100\,000 \div 100) \times (100\,000 \div 1000) =$

A) 100 000    B) 10 000    C) 10    D) 1

12. A

13. (the greatest odd factor of 2007) − (the least odd factor of 2007) =

A) 666    B) 668    C) 2004    D) 2006

13. D

14. 9 tenths + 9 hundredths = 9 thousandths + _?_

A) 0.9901    B) 0.981    C) 0.99    D) −0.901

14. B

15. The sum of a number and its reciprocal is *never*

A) 0    B) 2    C) positive    D) negative

15. A

*Go on to the next page* ⫸ **8**

28

Answers

16. A number that is divisible by 12 and 21 must be divisible by
    A) 28    B) 33    C) 36    D) 63

16. A

17. The average of $\frac{1}{2006}$ and $\frac{1}{2007}$ is equal to half of their
    A) sum    B) product    C) quotient    D) difference

17. A

18. $2^2 + 2^2 \times 2^2 + 2^2 \times 2^2 = 2^2 \times \underline{?}$
    A) 5    B) 6    C) 9    D) 16

18. C

19. Using my safari pencil, I found that the product of 120 and $\underline{?}$ is the square of an integer.
    A) 6    B) 10    C) 15    D) 30

19. D

20. The sum 0.1 + 0.01 + 0.001 is equal to $\underline{?}$ thousandths.
    A) 1    B) 11    C) 100    D) 111

20. D

21. Half of 0.5% is 5 times as large as
    A) 5%    B) 0.05%    C) 0.005%    D) 0.0005%

21. B

22. A candle 15 cm long, burning 0.5 cm/hr., takes $\underline{?}$ to burn halfway.
    A) 3.75 hours    B) 7.5 hours    C) 15 hours    D) 22.5 hours

22. C

23. Each of my bricks is a 1 by 2 by 3 rectangular solid. The sum of the areas of all 6 faces of a brick is
    A) 6    B) 11    C) 22    D) 36

23. C

24. $\dfrac{1}{1+\frac{1}{2}} = \dfrac{2}{?}$
    A) 1    B) 2    C) 3    D) 4

24. C

25. One-eighth of one-eighth = $\underline{?}$ of one-sixteenth
    A) one-sixteenth    B) one-fourth    C) one-third    D) one-half

25. B

26. The length of each side of a regular $\underline{?}$ is 20% of its perimeter.
    A) pentagon    B) hexagon    C) octagon    D) decagon

26. A

27. The number 30 is divisible by exactly $\underline{?}$ different whole numbers.
    A) 4    B) 6    C) 7    D) 8

27. D

28. What is the sum of 9 consecutive integers whose average is 10?
    A) 14    B) 19    C) 80    D) 90

28. D

29. If 4 CDs on sale cost the same as 3 CDs at full price, and 1 CD at full price costs $16, then how much do 9 CDs on sale cost?
    A) $144    B) $108    C) $72    D) $48

29. B

*Go on to the next page* ⇒ **8**

29

30. If a radius of the circle shown is 6, what is the area of the shaded region?

    A) $36-12\pi$   B) $144-12\pi$   C) $36-36\pi$   D) $144-36\pi$

    30.

31. Divide each of the 1999 numbers 1, 2, . . . , 1999 by 2. The average value of these 1999 quotients is

    A) 498.5       B) 499       C) 499.5       D) 500

    31.

32. If $a\triangle b$ means $b^{a}+(a\times b)$, then $3\triangle 4 =$

    A) 76    B) 84    C) 88    D) 93

    32.

33. In my piggy bank, 70% of the coins were dimes. Your piggy bank had three times as many coins as mine, and 75% of them were dimes. When the contents were combined, _?_ % of the coins were dimes.

    A) 72.5    B) 73.25    C) 73.75    D) 74

    33.

34. The least common multiple of 1, 2, 3, 4, 5, 6, 7, 8, 9, and 10 is

    A) 3 628 800    B) 7560    C) 2520    D) 1260

    34.

35. $1000^{1001} - 1000^{1000} =$

    A) 1000    B) $999^{1000}$    C) $1000\times 1001$    D) $999\times 1000^{1000}$

    35.

36. On my softball team, five of the player's numbers are primes. If the sum of these different primes is odd, this sum could be

    A) 39    B) 35    C) 27    D) 25

    36.

37. If the measures of the angles of a triangle are consecutive whole numbers, then the triangle must be

    A) scalene       B) obtuse
    C) isosceles     D) equilateral

    37.

38. If the square root of the perimeter of an equilateral triangle is 6, then the length of one side of this triangle is

    A) 8       B) 9       C) 12       D) 16

    38.

39. Of the following, which is divisible by 3?

    A) $10^{200} + 1$    B) $10^{200} + 2$    C) $10^{200} + 3$    D) $10^{200} + 4$

    39.

40. I need at least _?_ colors to paint each square so any squares that touch are colored differently.

    A) 2       B) 3       C) 4       D) 5

    40.

*The end of the contest*  8

# 2007-2008 Annual 8th Grade Contest

*Tuesday, February 19 or 26, 2008*

**8**

## Instructions

- **Time** You will have only *30 minutes* working time for this contest. You might be *unable* to finish all 40 questions in the time allowed.

- **Scores** Please remember that *this is a contest, not a test*—and there is no "passing" or "failing" score. Few students score as high as 30 points (75% correct). Students with half that, 15 points, *should be commended!*

- **Format and Point Value** This is a multiple-choice contest. Each answer is an A, B, C, or D. Write each answer in the *Answers* column to the right of each question. A correct answer is worth 1 point. Unanswered questions get no credit. You **may** use a calculator.

2 3 5 7 11 13 17 19 23 29

1. $10 + 70 + 20 + 60 + 30 + 50 = 80 \times \underline{?}$
   A) 3    B) 4    C) 6    D) 10

2. Of the following, which number is divisible by 2, 4, 8, and 16?
   A) 1624    B) 2461    C) 3218    D) 4816

3. $0.05 \times 0.01 = 0.5 \times \underline{?}$
   A) 01    B) 0.01    C) 0.001    D) 0.0001

4. In a triangle with perimeter 1, the average of the side-lengths is
   A) 1    B) $\frac{1}{6}$    C) $\frac{1}{3}$    D) 3

5. In a backyard game, if Fido scored 6 points every 12 minutes, then he scored 36 points in $\underline{?}$ hours.
   A) 1    B) 1.2    C) 1.5    D) 2

6. If Fido turned 14 years old today, then 100 months ago Fido was $\underline{?}$ years old.
   A) 4    B) 5    C) 6    D) 7

7. $\frac{1}{2} \times \frac{1}{2} + \frac{1}{2} \times \frac{1}{2} = \frac{1}{2} \times \underline{?}$
   A) $\frac{1}{8}$    B) $\frac{1}{4}$    C) $\frac{1}{2}$    D) 1

8. 5 hundredths + $\underline{?}$ = 100 thousandths
   A) $\frac{5}{100}$    B) $\frac{20}{100}$    C) $\frac{10}{100}$    D) $\frac{95}{1000}$

9. Of the following, which number is greater than 0.2008?
   A) 0.208    B) 0.20    C) 0.0208    D) 0.20008

10. How many of the angles in a triangle *must* be acute?
    A) 0    B) 1    C) 2    D) 3

11. $\frac{22+2}{2} + \frac{33+3}{3} + \frac{44+4}{4} =$
    A) 30+3    B) 33+3    C) 50+5    D) 55+5

12. Yogi hibernates $\frac{1}{3}$ of $\frac{3}{5}$ of every year. That's $\underline{?}$ % of every year.
    A) 80    B) 50    C) 40    D) 20

13. $3^2 + 6^2 + 9^2 = 3^2 \times \underline{?}$
    A) 5    B) 6    C) 14    D) 15

14. My 4-test average is 85. I need a $\underline{?}$ on my 5th test to average 88.
    A) 91    B) 96    C) 98    D) 100

15. By which of the following is 333 333 333 divisible?
    A) 11    B) 33    C) 111    D) 3333

16. If $4! = 4 \times 3 \times 2 \times 1$, then $\dfrac{2! \times 3! \times 4!}{2 \times 3 \times 4} =$

 A) 1     B) $4!$     C) $2! \times 3!$     D) $2 \times 3 \times 4$

16.

17. If $n$ is a positive integer, which is *never* divisible by 5?

 A) $n+5$     B) $3n+4$     C) $4n+3$     D) $5n+1$

17. D

18. Divide the sum of the first 1000 primes by 2. The remainder is

 A) 0     B) 1     C) 2     D) 3

18. B

19. Of the following, which is closest in value to its own reciprocal?

 A) 0.01     B) 0.1     C) 1.01     D) 1.1

19. C

20. Rabbit was asked "What time is it 1234 hours after midnight?" The right time was __?__, and it took Rabbit 1234 hours to get the right answer.

 A) 10 A.M.   B) noon   C) 10 P.M.   D) midnight

20. A

21. If a rectangle has perimeter 24, its area is at most

 A) 24     B) 25     C) 36     D) 144

21. C

22. No whole-number power of 3 has units' digit

 A) 1     B) 3     C) 6     D) 9

22. C

23. Which product has the same value as $999\,999^2 - 999\,999$?

 A) $1\,000\,000 \times 999\,998$     B) $999\,999 \times 999\,998$
 C) $999\,999 \times 1$     D) $999\,998 \times 999\,998$

23. B

24. If a rectangle's side-lengths are integers, its perimeter *must* be

 A) even   B) odd   C) prime   D) $> 4$

24. A

25. Of 40 kids, 24 sing, 16 play the drums, and 10 do neither. The ratio of the number who both sing and play the drums to the number who do neither is

 A) 1:4    B) 3:5    C) 4:5    D) 1:1

25. D

26. In a rodeo, there are 3 horses for every 2 bulls. These animals have 140 hooves altogether. How many bulls are in this rodeo?

 A) 35     B) 28     C) 21     D) 14

26. D

27. The reciprocal of $\dfrac{1}{2+\frac{1}{2}}$ is

 A) $\dfrac{2}{2+1}$   B) $\dfrac{2+1}{2}$   C) $\dfrac{2}{5}$   D) $\dfrac{5}{2}$

27. D

28. 2 ℓ of 2% fat milk + 3 ℓ of 3% fat milk = 5 ℓ of __?__ fat milk.

 A) 2.5%     B) 2.6%     C) 5%     D) 6%

28. A

29. If I *start with* 2, and begin to count by 3's, my 50th number will be

 A) 149     B) 150     C) 151     D) 152

29. A

30. If the product of 3 whole numbers is 100, their sum *cannot* be
    A) 14      B) 30      C) 52      D) 102

30. C

31. At most how many of the numbers 3, 5, 7, and 8 can be written as a sum of two *primes*?
    A) 3   B) 2   C) 1   D) none

31. B

32. Right now, Pat is 8 years younger than Lee. In 5 years, Pat will be half Lee's age. How old is Lee now, in years?
    A) 16   B) 11   C) 5   D) 3

32. B

33. 2% of 2% = 4% of ?
    A) 10      B) 1      C) 0.1      D) 0.01

33. B

34. A cube with surface area 24 is cut into 8 identical smaller cubes. What is the volume of each of the smaller cubes?
    A) 1      B) 3      C) 6      D) 8

34. A

35. If $a^3 \times b^4 \times c^5$ is negative, then ? *cannot* be negative.
    A) $a$      B) $c$      C) both $a$ & $c$   D) both $b$ & $c$

35. C

36. What is the greatest common factor of $4^8$ and $8^4$?
    A) $2^{64}$      B) $2^{16}$      C) $2^{12}$      D) $2^4$

36. B

37. One day, after Dad bought 3/5 of the fish that I caught, he gave away 1/4 of the fish he bought from me. What fraction of the fish that I caught did Dad keep?
    A) $\frac{3}{20}$   B) $\frac{4}{20}$   C) $\frac{7}{20}$   D) $\frac{9}{20}$

37.

38. If a 2-digit number is 5 times its digits' sum, then its digits' product is
    A) 10   B) 20   C) 30   D) 40

38. B

39. How many positive factors of 270 are multiples of 9?
    A) 7      B) 8      C) 29      D) 30

39. A

40. How many different paths start at square 1 and end at square 9 if the only two legal moves are moving down one square or moving to the right one square?
    A) 4      B) 6      C) 8      D) 12

40. D

*The end of the contest* **8**

# 2008-2009 Annual 8th Grade Contest

*Tuesday, February 17 or 24, 2009*

## Instructions

- **Time** You will have only *30 minutes* working time for this contest. You might be *unable* to finish all 40 questions in the time allowed.

- **Scores** Please remember that *this is a contest, not a test*—and there is no "passing" or "failing" score. Few students score as high as 30 points (75% correct). Students with half that, 15 points, *should be commended!*

- **Format and Point Value** This is a multiple-choice contest. Each answer is an A, B, C, or D. Write each answer in the *Answers* column to the right of each question. A correct answer is worth 1 point. Unanswered questions get no credit. You **may** use a calculator.

1. $2\sqrt{25} - 2\sqrt{16} =$
   A) 2    B) 3    C) 4    D) 6
   
   1. A

2. Of the following, which is most nearly equal to 0.55?
   A) 0.49    B) 0.509    C) 0.549    D) 0.6
   
   2. C

3. $1 + \frac{1}{3} + 2 + \frac{2}{3} + 3 + \frac{3}{3} =$
   A) 9    B) 8    C) 7    D) 6
   
   3. B

4. The greatest factor of $39 \times 49 \times 59$ is
   A) 9    B) prime    C) even    D) odd
   
   4. D

5. 99 hundredths − 99 thousandths =
   A) 0.891    B) 0.81    C) 0.01    D) 0.001
   
   5. A

6. Of the following fractions, which has the greatest reciprocal?
   A) $-\frac{3}{4}$    B) $\frac{5}{6}$    C) $-\frac{7}{8}$    D) $\frac{9}{10}$
   
   6. B

7. If filled identically, 8 dozen recycling containers might contain a total of _?_ items intended for recycling.
   A) 280   B) 284   C) 288   D) 292
   
   7. C

8. The month that occurs 4000 days after June 1 is
   A) April    B) May    C) June    D) July
   
   8.

9. 40% of 30% of 20% of 10% of 0% =
   A) 0    B) 100    C) 100%    D) 240 000%
   
   9. A

10. If a square's perimeter is three-fourths, its area is
    A) $\frac{3}{4}$   B) $\frac{3}{16}$   C) $\frac{9}{4}$   D) $\frac{9}{256}$
    
    10. D

11. I got paid for 8 hours of work at a victory party, but the host added 20% as my tip and gave me $120. My hourly wage, with no tip, was
    A) $10   B) $12   C) $12.50   D) $15
    
    11. C

12. 250% has the same value as
    A) $\frac{1}{4}$   B) $\frac{2}{5}$   C) $\frac{5}{2}$   D) 25
    
    12. C

13. The least common multiple of 11, 22, 33, and 44 is
    A) 66    B) 88    C) 99    D) 132
    
    13. D

14. The average of all the integers from −50 through 51, inclusive, is
    A) 0    B) 0.5    C) 1    D) 50
    
    14. B

2 4 8 16 32 64

248
3
144

15. I read twice as many pages each day as I read the day before. If I read my first 2 pages on Sunday, then I read my 100th page on
    A) Friday    B) Saturday    C) Monday    D) Tuesday

    15. A

16. After Al took 25% of my books and Ed took 50% of the remainder, only 30 of my books remained. How many books did Al take?
    A) 15    B) 20    C) 60    D) 80

    16. B

    36

17. Of the following, the largest is
    A) –0.1    B) $(-10)^3$    C) –100    D) $-\sqrt{100}$

    17. A

18. At most ? $2 \times 3 \times 6$ bricks fit into a $3 \times 6 \times 8$ space.
    A) 1.5    B) 2    C) 3    D) 4

    18.

19. ? is half as many minutes after 8:15 A.M. as before 3:45 P.M.
    A) 4 A.M.    B) 10:45 A.M.    C) 12 P.M.    D) 1:15 P.M.

    19. C

20. If the product of two consecutive whole numbers is 600, their sum is
    A) 48    B) 49    C) 50    D) 60

    20. B

21. If a circle's area is numerically 8 times its circumference, its radius is
    A) 2    B) 4    C) 8    D) 16

    21. D

22. The first ten numbers of a certain sequence are 1, 2, 2, 3, 3, 3, 4, 4, 4, and 4. The sum of the reciprocals of these ten numbers is
    A) 4    B) 3    C) 2    D) 1

    22. A

23. If 5 bowls = 2 cups, and 3 mugs = 4 bowls, then 8 cups =
    A) 20 mugs    B) 16 mugs    C) 15 mugs    D) 12 mugs

    23. C

24. How many different whole numbers are factors of both 24 and 124?
    A) 1    B) 2    C) 3    D) 4

    24. C

    1 2 4

25. My 10 flights cost $95 each. Your 20 flights cost $86 each. These 30 flights had an average cost of
    A) $89    B) $90    C) $91    D) $92

    25. A

    950
    1720
    2570

26. $(-1)^2 - (-1^2) =$
    A) –2    B) –1    C) 1    D) 2

    26.

27. $\sqrt{36}$ is half of
    A) $\sqrt{18}$    B) $\sqrt{72}$    C) $\sqrt{128}$    D) $\sqrt{144}$

    27. D

28. How many two-digit integers are twice the sum of their digits?
    A) 4    B) 2    C) 1    D) 0

    28.

29. A side of rectangle $R$ is 5 cm long, and one of its diagonals is 13 cm long. $R$'s area is _?_ cm².
    A) 34   B) 36   C) 60   D) 65

29. C

30. If a square's area is _?_ , its area numerically exceeds its perimeter.
    A) $\frac{\pi^2}{16}$   B) $\frac{16}{\pi^2}$   C) 4   D) $16\pi^2$

30. B

31. Of the following, _?_ has the largest power of 2 as a factor.
    A) $18^{36}$   B) $24^{30}$   C) $30^{24}$   D) $36^{18}$

31. B

32. In any isosceles triangle with a perimeter of 18, the length of the triangle's shortest side *cannot* be
    A) 4   B) 5   C) 6   D) 7

32. C

33. A circle can intersect a rectangle in at most _?_ points.
    A) 8   B) 7   C) 6   D) 4

33. A

34. The average measure of two of a right triangle's angles *could* be
    A) 30°   B) 40°   C) 50°   D) 90°

34. C

35. If $\left(\frac{a}{b}\right)^{-c}$ equals $\left(\frac{b}{a}\right)^c$, then $\left(\frac{2}{3}\right)^{-4} =$
    A) $\frac{81}{16}$   B) $\frac{16}{3}$   C) $\frac{3}{16}$   D) $\frac{2}{81}$

35. A

36. Divide a circle's area by the square of its circumference to get
    A) $\frac{1}{4\pi}$   B) $\frac{1}{4\pi^2}$   C) $\frac{1}{2\pi}$   D) $\frac{1}{4}$

36. A

37. Of the integers 1, 2, 3, ... , 2010, how many are divisible by at least one prime less than 2010?
    A) 2010   B) 2009
    C) 1004   D) 1005

37. B

38. The sum of 1024 fours is
    A) $4^4$   B) $4^5$   C) $4^6$   D) $4^7$

38. C

39. If I divide an integer by _?_ , the quotient *might be* equivalent to $\frac{24}{42}$ .
    A) 35   B) 45   C) 55   D) 65

39.

40. The sum of the first 50 of the first 200 positive integers is 1275. What is the sum of the last 50 of the first 200 positive integers?
    A) 6275   B) 6600   C) 8375   D) 8775

40.

*The end of the contest* ✍ **8**

# 2009-2010 Annual 8th Grade Contest

*Tuesday, February 16 or 23, 2010*

## Instructions

- **Time** Do *not* open this booklet until you are told by your teacher to begin. You might be *unable* to finish all 40 questions in the 30 minutes allowed.

- **Scores** Please remember that *this is a contest, and not a test*—there is no "passing" or "failing" score. Few students score as high as 30 points (75% correct). Students with half that, 15 points, *should be commended!*

- **Format, Point Value, & Eligibility** Every answer is an A, B, C, or D. Write answers in the *Answers* column. A correct answer is worth 1 point. Unanswered questions get no credit. You **may** use a calculator.

1. 2010 is *not* divisible by

   A) 2     B) 3     C) 5     D) 7

   1. D

2. 28% is equal to

   A) 2.8   B) $\frac{7}{25}$   C) 2800   D) $\frac{0.28}{100}$

   2. B

3. $\frac{4}{5} - \frac{3}{20} =$   $\frac{16}{20}\frac{3}{20} - \frac{13}{20}$

   A) $\frac{1}{15}$   B) $\frac{2}{15}$   C) $\frac{13}{20}$   D) $\frac{12}{65}$

   3. C

4. Al worked at the pool for 20% of the days in June. Al worked at the pool for _?_ days.

   A) 3     B) 6     C) 9     D) 12

   4. B

5. In a triangle with an acute angle, an obtuse angle, and a 60° angle, the obtuse angle could be

   A) 30°   B) 90°   C) 110°   D) 120°

   5. C

6. 0.8 =

   A) $\frac{4}{5}$   B) $\frac{5}{4}$   C) $\frac{1}{8}$   D) $\frac{8}{100}$

   6. A

7. If the time 6000 seconds ago was 10:00 AM, then what time is it now?

   A) 11:00 AM   B) 11:40 AM   C) 4:00 PM   D) 10:00 PM

   7. B

8. $2 - 13 - (-7) =$

   A) -18   B) -11   C) -8   D) -4

   8. D

9. The number that is 5 less than the square root of 144 is 5 more than

   A) $\sqrt{4}$   B) $\sqrt{25}$   C) $\sqrt{49}$   D) $\sqrt{134}$

   9. A

10. The product of a 3-digit integer and a 4-digit integer can have _?_ digits in all.

    A) 12    B) 8     C) 6     D) 5

    10. C

11. The hour and minute hands of a circular clock form a 60° angle at

    A) 2:00   B) 3:30   C) 6:00   D) 9:45

    11. A

12. The total number of primes between 24 and 42 is

    A) 3     B) 4     C) 5     D) 6

    12. B

13. $\frac{30 \times 25 \times 20 \times 15}{6 \times 5 \times 4 \times 3} =$

    A) $5^4$   B) $5^3$   C) $5^2$   D) $5^1$

    13. A

14. 30% of 40 is equal to 40% of

    A) 200   B) 120   C) 60    D) 30

    14. D

15. An equilateral triangle and a square have the same perimeter. If the length of a side of the triangle is 8, what is the area of the square?

    A) 16    B) 24    C) 36    D) 64

    15. C

*Go on to the next page* ))))▶ **8**

40

16. $\frac{1}{2} \div \frac{3}{8} =$

   A) $\frac{3}{16}$   B) $\frac{2}{5}$   C) $\frac{5}{8}$   D) $\frac{4}{3}$

16. D

---

17. $3^4 + 3^4 + 3^4 =$

   A) $3^5$   B) $9^4$   C) $3^{12}$   D) $9^{12}$

17. C

---

18. Eli's goal was to lift 3.5 kg during his workout, but he was able to lift only 3 kg. What fraction of his goal weight did Eli lift?

   A) $\frac{1}{7}$   B) $\frac{2}{3}$   C) $\frac{3}{4}$   D) $\frac{6}{7}$

18. D

---

19. Of the following, which is the largest?

   A) 0.02  B) $\frac{1}{20}$   C) 4%  D) $\frac{3}{100}$

19. B

---

20. The perimeter of square A is twice the perimeter of square B. The area of square A divided by the area of square B is

   A) $\frac{1}{2}$      B) 2      C) 4      D) 8

20. C

---

21. $13.25:1 = \underline{\ ?\ }:8$

   A) 53      B) 106      C) 122      D) 150

21. B

---

22. Of the following, which has the smallest reciprocal?

   A) $\frac{2}{5}$      B) $\frac{3}{7}$      C) $\frac{4}{3}$      D) $\frac{9}{4}$

22. D

---

23. At a certain school, there are 2 teachers for every 15 students. How many teachers are there if there are 165 students?

   A) 11      B) 13      C) 17      D) 22

23. D

---

24. The cost of 3 apples and 4 oranges is $4. If 2 oranges cost half as much as 1 apple, then the cost of 2 apples is

   A) $1.00 B) $1.50 C) $2.00 D) $2.50

24. C

---

25. $10^{2010} - 1$ is divisible by
   A) 10    B) 11    C) 12    D) 15

25. B

---

26. 250% of $\underline{\ ?\ }$ is 30.

   A) 12    B) 45    C) 70    D) 75

26. A

---

27. $2.01 \times 10^{2009} = 2010 \times \underline{\ ?\ }$

   A) $10^{2006}$      B) $10^{2007}$      C) $10^{2012}$      D) $10^{2013}$

27. A

---

28. How many prime numbers are divisible by 2?

   A) zero      B) one      C) three      D) ten

28. B

---

29. As shown, two small circles pass through opposite endpoints of a diameter of a large circle and touch once at its center. If the large circle's area is $16\pi$, then the shaded region's area is

   A) $4\pi$    B) $5\pi$    C) $6\pi$    D) $8\pi$

29. D

---

*Go on to the next page* )))➡ **8**

30. _?_ is *not* the sum of the squares of 3 integers.

A) 14   B) 21   C) 28   D) 35

30. C

31. If 9 months ago was January 1, then 90 months from now it will be

A) March   B) April   C) May   D) June

31. B

32. The sum of all positive integral factors of 32 is

A) 5   B) 30   C) 32   D) 63

32. D

33. Jim has 12 socks: 4 red, 4 black, and 4 blue. Choosing in the dark, he wants at least one matching pair of socks that are *not* red. If he does not know what color socks he is choosing, then he must choose at least _?_ socks to be sure he has a matching pair.

A) 2   B) 3   C) 6   D) 7

33. D

34. A number between 0 and 1 is multiplied by a number between 1 and 2. Which of the following could *not* be the product?

A) 0.25   B) 1   C) 1.75   D) 2.25

34. D

35. If A ◊ B means (A + B) × (B − A), then 1 ◊ (2 ◊ 3) =

A) 35   B) 24   C) 2   D) 0

35. B

36. _?_ could be the sum of 6 consecutive odd integers.

A) 108   B) 111   C) 333   D) 345

36. A

37. I count the petals on my 20 flowers. The first 8 average 24 petals each. The next 12 average 34 petals each. What is the average number of petals on all 20 flowers?

A) 28   B) 29   C) 30   D) 31

37. C

38. If 28 is a factor of the square of an integer, then another factor of the same square must be

A) 784   B) 49   C) 20   D) 12

38. D

39. On July 1, Harry sold $7 worth of lemonade. On July 2, he sold $10 worth. Each day after that he sold $3 more than he sold the day before. When did he first sell $100 worth in a single day?

A) July 30   B) July 31   C) August 1   D) August 2

39. C

40. My number is the square of an integer, the cube of an integer, and greater than 1. The least possible total number of positive divisors of my number is

A) 7   B) 6   C) 5   D) 4

40. B

*The end of the contest* ☞ **8**

**Visit our Web site at http://www.mathleague.com**

Solutions on Page 107 • Answers on Page 146

42

# 2010-2011 Annual 8th Grade Contest

*Tuesday, February 15 or 22, 2011*

**8**

## Instructions

- **Time** Do *not* open this booklet until you are told by your teacher to begin. You might be *unable* to finish all 35 questions in the 30 minutes allowed.

- **Scores** Please remember that *this is a contest, and not a test*—there is no "passing" or "failing" score. Few students score as high as 28 points (80% correct). Students with half that, 14 points, *should be commended!*

- **Format, Point Value, & Eligibility** Every answer is an A, B, C, or D. Write answers in the *Answers* column. A correct answer is worth 1 point. Unanswered questions get no credit. You **may** use a calculator.

1. I have \$222 but I need \$2011. If my friend loans me \$789, how much more do I need?

   A) \$0  B) \$100  C) \$1000  D) \$1111

   1. C

2. Prospector Al finds a gold nugget weighing 500 grams. In kg, two of these would weigh

   A) 1 kg B) 10 kg C) 1000 kg D) 10 000 kg

   2. A

3. $6 + 8 \times 0 + 10 \times 2 - 12 \times 0 =$

   A) 0   B) 8   C) 26   D) 38

   3. C

4. What is the probability that a randomly chosen positive integer less than 20 will be a multiple of 4?

   A) $\frac{4}{19}$   B) $\frac{1}{5}$   C) $\frac{5}{19}$   D) $\frac{1}{4}$

   4. A

5. The square of an integer minus the cube of the same integer is *never*

   A) positive   B) negative   C) even   D) odd

   5. D

6. Which of the following is *not* the product of two prime numbers?

   A) 85   B) 94   C) 119   D) 127

   6. D

7. $10 \times 10 \times 5 \times 20 \times 4 \times 25 \times 2 \times 50 \times 100 = \underline{\ ?\ } \times 100$

   A) 4   B) 5   C) $100^4$   D) $100^5$

   7. C

8. The ratio of the number of dimes in \$100 to the number of quarters in \$200 is

   A) 5:4   B) 4:5   C) 5:2   D) 2:5

   8. A

9. My three friends and I divide the cost of a restaurant dinner equally. If the cost was \$60 after a 20% tip was added, what was the cost for each of us without the tip?

   A) \$12   B) \$12.50   C) \$16   D) \$16.67

   9. B

10. It takes 12 hours for 18 workers to build a wall. At this rate, how many hours would it take 12 workers to build the wall?

    A) 6   B) 8   C) 14   D) 18

    10. D

11. Midway between 10:59 PM today and 11:01 PM tomorrow is tomorrow at

    A) 10 AM B) 11 AM C) 10 PM D) 11 PM

    11. B

12. Which of the following has the most factors of 5?

    A) 125   B) 500   C) 625   D) 750

    12. C

13. Eight hundredths is what percent of four thousandths?

    A) 5%   B) 20%   C) 500%   D) 2000%

    13. D

44          *Go on to the next page* ))) **8**

**Answers**

14. Mr. B. Loon has 2 fancy balloons for every 7 plain ones. He has a total of 621 balloons. How many are fancy?

   A) 138    B) 183    C) 207    D) 483

14. A

15. $\dfrac{\frac{1}{2}}{\frac{1}{3}+\frac{1}{4}}=$

   A) $\dfrac{4}{7}$    B) $\dfrac{6}{7}$    C) $\dfrac{7}{6}$    D) $\dfrac{7}{4}$

15. B

16. How many digits are in the decimal representation of the product of $2^5$ and $10^{52}$?

   A) 54    B) 55    C) 56    D) 57

16. A

17. I wrote the integers from 1 to 100 in order. The 50th digit I wrote was

   A) 0    B) 3    C) 4    D) 9

17. B

18. Which of the following has the fewest different prime factors?

   A) 30    B) 32    C) 34    D) 36

18. B

19. The difference between an angle of an equilateral triangle and an angle of an isosceles right triangle could be

   A) 15°    B) 45°    C) 60°    D) 75°

19. A

20. What is the least common multiple of $4^8$ and $8^4$?

   A) $4^4$    B) $4^8$    C) $8^4$    D) $8^8$

20. C

21. The sum of the digits of a prime number greater than 9 *cannot* be

   A) 2    B) 3    C) 4    D) 5

21. B

22. The number of seconds in an hour divided by the number of minutes in an hour is

   A) 5    B) 12    C) 60    D) 1440

22. C

23. My current age will be tripled 16 years after it's doubled. My current age is

   A) 8    B) 12    C) 16    D) 32

23. C

24. The average of 10 positive integers is 10. The greatest of them *cannot* be

   A) 10    B) 50    C) 90    D) 92

24. A

25. If each toboggan holds a prime number of riders, including at least 2 children and 1 adult, which of the following could be the number of riders on 21 toboggans?

   A) 110    B) 112    C) 121    D) 122

25. C

*Go on to the next page* ⟩⟩⟩➡ **8**

6 80

26. Each beaver in a colony of 20 beavers cuts 14 logs for a dam. Each beaver in another colony of 40 beavers cuts 20 logs. If the two colonies are combined, what is the average number of logs cut per beaver?

A) 18    B) 17    C) 16    D) 15

26. A

27. To decrease the area of a circle by 75%, I must decrease its radius by

A) 25%   B) 50%   C) 60%   D) 75%

27. B

28. The product of $\sqrt{2}$ and the reciprocal of the reciprocal of $\sqrt{2}$ is

A) $\sqrt{2}$      B) 2      C) $2\sqrt{2}$      D) 4

28. B

29. A store has 3 melons for every 8 apples, and 5 apples for every 9 pears. If there are 600 melons, how many pears are there?

A) 600      B) 1320      C) 1440      D) 2880

29. D

30. If $r \blacklozenge s$ means $r^2 - 2s$, what is the value of $3 \blacklozenge (4 \blacklozenge 5)$?

A) -9      B) -3      C) 0      D) 3

30. A

31. Jo had four times as many dimes as pennies in her pocket. She spent two dimes and got two pennies in change. Now she has three times as many dimes as pennies. Jo started with _?_ dimes.

A) 8      B) 12      C) 28      D) 32

31. D

32. The base of an isosceles triangle is the diameter of a semicircle, as shown. If the radius of the semicircle is 2, and the area of the entire figure is $2\pi + 16$, what is the greatest possible distance between two points of the figure?

A) 10    B) 12    C) 14    D) 16

32. A

33. Pat's drawer is 1 m wide, 2 m long, and 0.5 m deep. What is its capacity in cubic cm?

A) 0.01   B) 100   C) 10 000   D) 1 000 000

33. B

34. From one end of River Road to the other, house numbers always increase. House numbers on any 2 adjacent houses differ by the same amount. The 11th number is 2011. The 31st is 2131. The 1st number is

A) 1901   B) 1945   C) 1951   D) 1968

34. C

35. Of the positive integers with an odd number of positive factors, how many are less than 2011?

A) 1    B) 21    C) 32    D) 44

35. B

*The end of the contest* ☞ **8**

# Algebra Course 1 Contests

## 2006-2007 through 2010-2011

# 2006-2007 Annual Algebra Course 1 Contest

*Spring, 2007*

## Instructions

- **Time** You will have only *30 minutes* working time for this contest. You might be *unable* to finish all 30 questions in the time allowed.

- **Scores** Please remember that *this is a contest, not a test*—and there is no "passing" or "failing" score. Few students score as high as 24 points (80% correct). Students with half that, 12 points, *deserve commendation!*

- **Format and Point Value** This is a multiple-choice contest. Each answer is an A, B, C, or D. Write each answer in the *Answer Column* to the right of each question. A correct answer is worth 1 point. Unanswered questions get no credit. You **may** use a calculator.

Answer Column

1. $(3+6)^2 = 3^2+6^2+$ ?

    A) 0        B) 9        C) 18        D) 36

1. D

2. When $x = 2007$, $2(x+0) + 0(x+7) =$

    A) 2    B) 2007    C) 4007    D) 4014

2. D

3. On a treadmill, I was able to compute that $(-2)(-3)(-4)(-5) = (-3)(-4)($ ? $)$.

    A) −20    B) −10    C) 10    D) 20

3. C

4. If $\frac{a}{b} = -2$, then $2b =$

    A) $-a$    B) $-4a$    C) $a$    D) $4a$

4. A

5. If $x+1 = y$, then $x^2+2x+1 =$

    A) $2y+1$      B) $y^2$      C) $y^2+1$      D) $y^2+2x$

5. B

6. In which interval does $x^2$ attain its least value?

    A) $-3 < x < -1$   B) $-2 < x < 1$   C) $0 < x < 1$   D) $1 < x < 2$

6. B

7. $10x^2-5x+(-3x)-(-2x^2) =$

    A) $12x^2-8x$   B) $12x^2-2x$   C) $8x^2-8x$   D) $8x^2-2x$

7. B

8. In a 2 hour parade, 40% of the drummers passed me in the 1st hour. If 45 drummers passed me in the 2nd hour, then the total number of drummers in the parade was

    A) 63    B) 70    C) 72    D) 75

8. D

9. Whenever $x+10$ is odd, $x+5$ is always

    A) odd      B) even      C) prime      D) positive

9. B

10. What is the product of the roots of $(x-2)(x-3) = 0$?

    A) 0      B) −5      C) −6      D) 6

10. D

11. What is the greatest common factor of $x^2+2x+1$ and $x^2+3x+2$?

    A) 2      B) $x^2$      C) $x+1$      D) $x+2$

11. C

*Go on to the next page* ⟾ **A**

12. The slopes of 2 parallel lines can have a product of $\frac{1}{4}$ and a sum of

A) $\frac{1}{16}$   B) $\frac{1}{4}$   C) $\frac{1}{2}$   D) 1

12. D

13. The number of condiments on my sandwich equals the number of composite factors of 30 = 2×3×5. That number is

A) 3   B) 4   C) 6   D) 7

13. B

14. For every negative value of $x$, $|x-2|$ =

A) $|x|+2$   B) $|x|-2$   C) $2-|x|$   D) $x+2$

14. A

15. _?_ different **integers** satisfy $(x^2-4)(x^2-16)(x^2-20)(x^2-36) = 0$.

A) 3   B) 4   C) 6   D) 8

15. D   C

16. If both (1,3) and (6,3) lie on the line $y = mx+b$, then $mb$ =

A) 6   B) 5   C) 3   D) 0

16. D

17. Which *cannot* be written as a sum of the squares of two integers?

A) 17   B) 18   C) 19   D) 20

17. C

18. If $x = 1000$, then 0.5% of $x$ =

A) 0.5   B) 5   C) 50   D) 500

18. B

19. The Cafeteria sign is a rectangle whose width is $x^2-5x-6$. If the sign's length is $\dfrac{1}{x^2-2x-3}$, its area is

A) 2   B) $\dfrac{x-2}{x+1}$

C) $\dfrac{x-6}{x-3}$   D) $\dfrac{5x+6}{2x+3}$

19. C

20. How many different negative integers satisfy $x^2 < 2007$?

A) 44   B) 45   C) 88   D) 89

20. A

21. _?_ is *not* the product of two binomials with integer coefficients.

A) $x^2+2xy+y^2$   B) $x^2-2xy+y^2$   C) $x^2-2xy-y^2$   D) $x^2-y^2$

21. D

*Go on to the next page* ⟫ **A**

51

*(handwritten at top: $(x+y)(x+y)(x+y)$  $3x^2+6xy+$  $x^2+xy+x^2+xy+ yx+y^2+yx+y^2+x^2+xy+y^2+yy$)*

22. $\sqrt{x^{64}} \div \sqrt{x^4} =$

    A) $x^4$    B) $x^6$    C) $x^{16}$    D) $x^{30}$

    22. **A**

23. The number of times that I danced with a star is $(x+3)^2-(x-3)^2$. If $x$ is an integer greater than 3, how many times did I dance with a star?

    A) 0    B) 3x    C) 12x    D) 18

    23. **C**

24. If $x > 0$ and $x^2+x^2+x^2+x^2=x^4$, then $x^4+x^4+x^4+x^4$ must equal

    A) $x^8$    B) $4x^6$    C) $6x^4$    D) $8x^3$

    24. **A**

25. The sum of the coefficients of the 4-term expansion of $(x+y)^3$ is

    A) 8    B) 6    C) 4    D) 3

    25. **A**

26. How many factors of $10^5$ are squares of integers?

    A) 4    B) 9    C) 16    D) 25

    26. **B**

27. The graph of __?__ crosses the x-axis in 2 different points.

    A) $y = x^2-64$    B) $y = x^2$
    C) $y = x^2+10x+25$    D) $y = x^2-24x+144$

    27. **A**

28. If line $\ell$ has slope $\dfrac{1}{1+\dfrac{1}{x}}$, what is the slope of any line that is perpendicular to line $\ell$?

    A) $-x$    B) $-\dfrac{x}{x+1}$    C) $-\dfrac{x+1}{x}$    D) $-\dfrac{1}{x+1}$

    28. **C**

29. I added $10^{2006}$ and $10^{2007}$. Their sum is a __?__-digit number.

    A) 2007    B) 2008    C) 4013    D) 4015

    29. **A**

30. If $x+y = 6$ and $xy = 4$, then $x^2+y^2 =$

    A) 20    B) 28    C) 32    D) 36

    30. **B**

*The end of the contest* **A**

# 2007-2008 Annual Algebra Course 1 Contest

*Spring, 2008*

## Instructions

- **Time** You will have only *30 minutes* working time for this contest. You might be *unable* to finish all 30 questions in the time allowed.

- **Scores** Please remember that *this is a contest, not a test*—and there is no "passing" or "failing" score. Few students score as high as 24 points (80% correct). Students with half that, 12 points, *deserve commendation!*

- **Format and Point Value** This is a multiple-choice contest. Each answer is an A, B, C, or D. Write each answer in the *Answer Column* to the right of each question. A correct answer is worth 1 point. Unanswered questions get no credit. You **may** use a calculator.

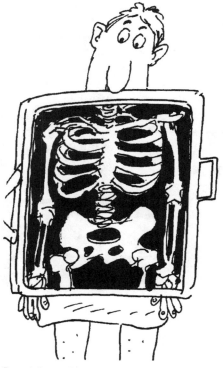

1. $(2 + 0 + 0 + 8)^0$

    A) 0  B) 1  C) 4  D) 9

    1.

2. If $x^2 = 10$, then $(x+1)(x-1) =$

    A) 99  B) 11  C) 9  D) –1

    2.

3. Joe painted 50% more houses in 2007 than in 2006. Joe painted 36 houses in 2007. How many houses did Joe paint in 2006?

    A) 12  B) 18  C) 24  D) 27

    3.

4. Of the following fractions, which one is reducible?

    A) $\frac{x^2-1}{x+1}$  B) $\frac{x^2-2}{x+2}$

    C) $\frac{x^2-3}{x+3}$  D) $\frac{x^2-4}{x+4}$

    4.

5. How many positive integers are factors of $2^{2008}$?

    A) 1  B) 2007  C) 2008  D) 2009

    5.

6. If $100x+100y = (x+y)^2$, and $x+y \neq 0$, what is the value of $x+y$?

    A) 10  B) 100  C) $100+100$  D) $100^2$

    6.

7. $(x+1)^2 - (x-1)^2 =$

    A) 0  B) $-2x^2$  C) $2x$  D) $4x$

    7.

8. If $a \geq b > 0$, then _?_ different pairs of integers $(a,b)$ satisfy $\frac{a}{8} = \frac{8}{b}$.

    A) 7  B) 6  C) 4  D) 3

    8.

9. If $50 < x < 150$ and $\sqrt{x}$ is prime, then the sum of the digits of $x$ is

    A) 2  B) 4  C) 11  D) 121

    9.

10. $x^2+5x-6$ is divisible by

    A) $x-6$  B) $x-3$  C) $x-2$  D) $x-1$

    10.

11. If $|x-y| > x-y$, which of the following *must* be true?

    A) $y > x$  B) $x > y$  C) $y < 0$  D) $x > 0$

    11.

*Go on to the next page* ⫸ **A**

12. What is the smallest of 16 consecutive integers whose sum is 8?

A) –6    B) –7    C) –8    D) –16

12.

13. The $y$-intercept of $y = x+1$ equals the $x$-intercept of

A) $y = x-1$    B) $y = x$
C) $y = x+1$    D) $y = x^2$

13.

14. If $(x-1)(x+1)(x-2)(x+2)$ is expanded, and like terms are then combined, the result has exactly ? terms.

A) 3    B) 6    C) 9    D) 12

14.

15. $x[x(x^2)^2]^2 =$

A) $x^8$    B) $x^9$    C) $x^{10}$    D) $x^{11}$

15.

16. For how many different integers $b$ is $x^2+bx+12$ factorable?

A) 6    B) 5    C) 4    D) 3

16.

17. The line ? passes through *neither* quadrant I nor quadrant III.

A) $y = 2x+4$    B) $y = -2x+4$    C) $y = 4x$    D) $y = -4x$

17.

18. What are all real values of $x$ for which $\frac{1}{4x^4-4}$ is undefined?

A) 0    B) $\pm 1$    C) $\pm 2$    D) $\pm 4$

18.

19. $999\,999^{6x} \div 999\,999^{3x} =$

A) $999\,999^2$    B) $999\,999^{2x}$    C) $999\,999^{3x}$    D) $999\,999^3$

19.

20. Of the following lines, which has the greatest slope?

A) $x+y = 2$    B) $x-y = 2$    C) $2x+y = 2$    D) $2x-y = 2$

20.

21. When we played them, our score was $\frac{2.2}{3.3+4.4} + 1.1$. If we tied them, then their score would have been $\frac{22}{33+44} + $ ?.

A) 0.011    B) 0.11    C) 1.1    D) 11

21.

22. If $-1 < x < 0$, then

A) $x > x^2$    B) $x > x^3$    C) $x^2 > x^3$    D) $x^3 > x^2$

22.

*Go on to the next page* ⇒ **A**

55

23. Each of the $p$ pigs in my truck wore a different integer from 1 to $p$. What is the sum of the numbers worn by all $p$ pigs?

A) $\frac{1}{2}p(1+p)$    B) $p(1+p)$

C) $\frac{1}{2}(1+p)$    D) $\frac{1}{2}p(p-1)$

23.

24. $2\sqrt{8} + 8\sqrt{2} =$

A) $\sqrt{256}$    B) $\sqrt{288}$
C) $\sqrt{384}$    D) $\sqrt{512}$

24.

25. How many ordered pairs of integers $(x,y)$ satisfy $x^2+y^2 = 2$?

A) 2    B) 4    C) 6    D) 8

25.

26. For every even integer $x > 0$, the sum $-x^x + (-x)^x$ is

A) positive    B) negative    C) zero    D) undefined

26.

27. If $x+\pi = 2$, then $x^2+3\pi x+2\pi^2 =$

A) $2+2\pi$    B) $4+\pi$    C) $2+\pi$    D) $4+2\pi$

27.

28. I saw my bones on an x-ray machine's rectangular screen. The sides of that screen had integral lengths, and the ratio of its length to its width was 5:4. The area of the screen could be

A) 200    B) 600    C) 4000    D) 8000

28.

29. I averaged $\frac{1}{k}$ km/hr on a 2 km run. If I ran the 1st km at $k$ km/hr., then I must have run the 2nd km at ? km/hr.

A) $\frac{k^2-2}{k}$    B) $\frac{2-k^2}{k}$    C) $\frac{k}{2k^2-1}$    D) $\frac{k^2+1}{2k}$

29.

30. What is the 2008th term of the sequence $\frac{\sqrt{2}}{2}, \frac{\sqrt{4}}{4}, \frac{\sqrt{8}}{8}, \ldots, \frac{\sqrt{2^n}}{2^n}$?

A) $\frac{1}{2^{1004}}$    B) $\frac{1}{1004}$    C) $\frac{1}{2^{2008}}$    D) $\frac{1}{2008}$

30.

*The end of the contest* ✍ **A**

**Visit our Web site at http://www.mathleague.com**

Solutions on Page 121 • Answers on Page 149

56

# 2008-2009 Annual Algebra Course 1 Contest

*Spring, 2009*

## Instructions

- **Time**  You will have only *30 minutes* working time for this contest. You might be *unable* to finish all 30 questions in the time allowed.

- **Scores**  Please remember that *this is a contest, and not a test*—there is no "passing" or "failing" score. Few students score as high as 24 points (80% correct). Students with half that, 12 points, *deserve commendation!*

- **Format and Point Value**  This is a multiple-choice contest. Each answer is an A, B, C, or D. Write each answer in the *Answer Column* to the right of each question. A correct answer is worth 1 point. Unanswered questions receive no credit. You **may** use a calculator.

Answers

1. If $m = 2$, $a = 3$, $t = 5$, and $m + a + t + h = 22$, what is the value of $h$?

   A) 4    B) 7    C) 9    D) 12

**1.** D

2. The number of years in 21 centuries is the same as the number of years in __?__ decades.

   A) 12    B) 21    C) 210    D) 2100

**2.** C

3. $(x + 1) + (2x + 3) + (4x + 5) + (6x + 7) =$

   A) $12x + 16$    B) $13x + 16$
   C) $13x + 12$    D) $16x + 13$

**3.** B

4. The square of a certain number is greater than the cube of the number but less than the number itself. Which of the following could be the number?

   A) $\dfrac{1}{3}$    B) $-\dfrac{1}{3}$    C) 3    D) –3

**4.** B

5. $(x + 1)(x - 1) - (x + 2)(x - 2) =$

   A) –5    B) 3    C) $2x^2 - 5$    D) $2x^2 + 3$

$x^2 - 1 - x^2 - 4$

**5.** A

6. If $x^2 = 3$, then $x^4 - 3 =$

   A) 0    B) 3    C) 6    D) 9

**6.** C

7. Which is 200% greater than $x$?

   A) $x + 200$    B) $x + 300$    C) $2x$    D) $3x$

**7.** D

8. If $x = y$, then $(x + y)(x - y) =$

   A) 0    B) $x^2 - 2xy + y^2$    C) $x^2 + 2xy - y^2$    D) $x^2 + y^2$

**8.** A

9. $cd^2 = c \div$ __?__

   A) $\dfrac{1}{d}$    B) $\dfrac{1}{d^2}$    C) $\dfrac{d^2}{c}$    D) $\dfrac{c}{d^2}$

**9.** B

10. If $p$ is a prime number greater than 2009, then __?__ cannot be a prime number.

    A) $p + 10$    B) $p - 10$    C) $10p$    D) $p + 1000$

**10.** C

11. The addresses of the five houses in the town where Alice lives are consecutive integers that add up to 10055. What is the *greatest* of these five integers?

    A) 2009    B) 2010    C) 2012    D) 2013

**11.** A

Go on to the next page ))))➡ **A**

12. Ben runs $x$ **m** per minute. How many **km** does he run in $y$ minutes?

A) $\dfrac{x}{1000y}$    B) $\dfrac{y}{1000x}$    C) $\dfrac{xy}{1000}$    D) $\dfrac{1000}{xy}$

12. C

13. The greatest common factor of $5^{555}$ and $5^{777}$ is

A) 5    B) $5^{111}$    C) $5^{555}$    D) $5^{777}$

13. D

14. If $y = 2008^{2008}$, which of the following has the greatest value?

A) $y^{2008}$    B) $\sqrt{y^{4008}}$    C) $(y^{200})^8$    D) $y^2 y^{1008}$

14. D

15. If the $x$- and $y$-intercepts of a line are both positive numbers, then the slope of the line must be

A) positive    B) negative    C) 0    D) undefined

15. B

16. $2^{100} \times (-2)^{101} \div 2^{202} =$

A) 2    B) –2    C) $\dfrac{1}{2}$    D) $-\dfrac{1}{2}$

16. D

17. The sum of all values of $x$ that satisfy $(x - 11)^2 - 1 = 0$ is

A) –2    B) 1    C) 2    D) 22

17. D

18. If $x = \underline{\ ?\ }$, then $x^3$ is the square of an integer.

A) 20    B) 25    C) 30    D) 35

18. B

19. If line $\ell$ has an integral slope, then the sum of $\ell$'s slope and the slope of a line perpendicular to $\ell$ could be

A) $\dfrac{3}{2}$    B) $\dfrac{7}{3}$    C) $\dfrac{9}{5}$    D) $\dfrac{8}{7}$

19. A

20. Charlie does yard work for the math teacher next door. The teacher agrees to pay Charlie $5, plus $$x$ per hour, where $x$ is equal to the sum of all integers $t$ for which $t^2 \le 40$. It takes Charlie 6 hours to finish the job, so how much will he be paid?

A) $5    B) $26    C) $131    D) $156

20. C

21. A rectangle is drawn in the $xy$-plane such that one of its diagonals has endpoints at $(2,6)$ and at $(14,1)$. If a third vertex of the rectangle is at $(x,6)$, the value of $x$ is

A) 1    B) 2    C) 6    D) 14

21. D

Go on to the next page ⟫⟫ **A**

59

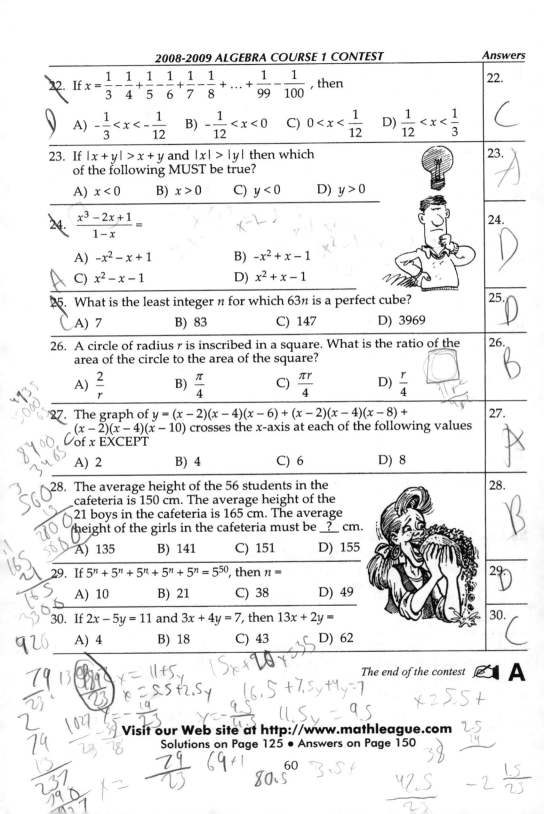

22. If $x = \dfrac{1}{3} - \dfrac{1}{4} + \dfrac{1}{5} - \dfrac{1}{6} + \dfrac{1}{7} - \dfrac{1}{8} + \ldots + \dfrac{1}{99} - \dfrac{1}{100}$ , then

    A) $-\dfrac{1}{3} < x < -\dfrac{1}{12}$    B) $-\dfrac{1}{12} < x < 0$    C) $0 < x < \dfrac{1}{12}$    D) $\dfrac{1}{12} < x < \dfrac{1}{3}$

22. C

23. If $|x+y| > x+y$ and $|x| > |y|$ then which of the following MUST be true?

    A) $x < 0$     B) $x > 0$     C) $y < 0$     D) $y > 0$

23. A

24. $\dfrac{x^3 - 2x + 1}{1 - x} =$

    A) $-x^2 - x + 1$       B) $-x^2 + x - 1$

    C) $x^2 - x - 1$        D) $x^2 + x - 1$

24. D

25. What is the least integer $n$ for which $63n$ is a perfect cube?

    A) 7        B) 83        C) 147        D) 3969

25. D

26. A circle of radius $r$ is inscribed in a square. What is the ratio of the area of the circle to the area of the square?

    A) $\dfrac{2}{r}$       B) $\dfrac{\pi}{4}$       C) $\dfrac{\pi r}{4}$       D) $\dfrac{r}{4}$

26. B

27. The graph of $y = (x-2)(x-4)(x-6) + (x-2)(x-4)(x-8) + (x-2)(x-4)(x-10)$ crosses the $x$-axis at each of the following values of $x$ EXCEPT

    A) 2        B) 4        C) 6        D) 8

27. A

28. The average height of the 56 students in the cafeteria is 150 cm. The average height of the 21 boys in the cafeteria is 165 cm. The average height of the girls in the cafeteria must be __?__ cm.

    A) 135       B) 141       C) 151       D) 155

28. B

29. If $5^n + 5^n + 5^n + 5^n + 5^n = 5^{50}$, then $n =$

    A) 10       B) 21       C) 38       D) 49

29. D

30. If $2x - 5y = 11$ and $3x + 4y = 7$, then $13x + 2y =$

    A) 4       B) 18       C) 43       D) 62

30. C

*The end of the contest* **A**

# 2009-2010 Annual Algebra Course 1 Contest

*Spring, 2010*

## Instructions

- **Time** Do *not* open this booklet until you are told by your teacher to begin. You will have only *30 minutes* working time for this contest. You might be *unable* to finish all 30 questions in the time allowed.

- **Scores** Please remember that *this is a contest, and not a test*—there is no "passing" or "failing" score. Few students score as high as 24 points (80% correct). Students with half that, 12 points, *should be commended!*

- **Format and Point Value** This is a multiple-choice contest. Each answer is an A, B, C, or D. Write each answer in the *Answer Column* to the right of each question. A correct answer is worth 1 point. Unanswered questions receive no credit. You **may** use a calculator.

1.  What is the value of $b + e + a + r$
    if $b = 1$, $e = 2b$, and $a = 3e$, and $r = 4a$?

    A) 4        B) 10        C) 24        D) 33

1.

2.  Of the following, which has the greatest value
    when $x = 1$?

    A) $x - 1$        B) $x^2 - 2$        C) $x^3 - 3$        D) $x^4 - 4$

2.

3.  $(-4)^2(-3)^0(-2)^1(-1)^0 =$

    A) $-32$        B) 0        C) 16        D) 32

3.

4.  __?__ is a factor of $x^4 - 16$.

    A) $x + 4$        B) $x - 4$        C) $x + 2$        D) $x - 1$

4.

5.  For which of the following values of $x$ is the value of $x^{-2010}$
    greatest?

    A) 100        B) 200        C) 300        D) 400

5.

6.  If $300x = 450 - 300y$, then $x + y =$

    A) $-3$        B) 1.5        C) 15        D) 30

6.

7.  $(2x^4 + 4x^2) + (3x^4 - 5x^2) - (4x^4 - 6x^2) =$

    A) $x^4 - 7x^2$        B) $x^4 + 5x^2$        C) $x^4 - 3x^2$        D) $x^4 + x^2$

7.

8.  If $x > 0$, then the additive inverse of $x$ divided by the reciprocal of $x$
    equals

    A) the square of $x$        B) the reciprocal of the square of $x$
    C) the square root of $x$        D) the additive inverse of the square of $x$

8.

9.  $((a^4)^3)^2 =$

    A) $a^9$        B) $a^{24}$        C) $a^{36}$        D) $a^{64}$

9.

10. One day, $c$ children ate $g$ giant ice cream cones.
    If $g - c = 2$ and $g^2 - c^2 = 20$, then $g =$

    A) 6        B) 8        C) 10        D) 12

10.

11. $x\%$ of $y\%$ of $100\,000 =$

    A) $x + y$        B) $xy$        C) $\dfrac{x + y}{10}$        D) $10xy$

11.

12. $\sqrt{2} + \sqrt{4} + \sqrt{8} + \sqrt{16} =$

    A) $\sqrt{30}$        B) $6 + 3\sqrt{2}$        C) $10\sqrt{2}$        D) $3 + 4\sqrt{2}$

12.

*Go on to the next page* ))))⮕ **A**

13. My teacher's blackboard is in the shape of a rectangle. I add its area to its perimeter, then subtract twice its length, then divide by its width, then subtract 2 from the quotient. The result is the rectangle's

A) length  B) width  C) diagonal  D) area

13.

14. If the equation of line $\ell$ is $39x + 54y = 101$, then an equation of a line perpendicular to $\ell$ is

A) $54x + 39y = 16$

B) $36x + 26y = 56$

C) $54x - 39y = 1$

D) $39x - 54y = 101$

14.

15. If $x > 5$, then $\dfrac{x-1}{x-2} \times \dfrac{x-2}{x-3} \times \dfrac{x-3}{x-4} \times \dfrac{x-4}{x-5} =$

A) $x^4$    B) $(x-1)^4$    C) $(x-5)^{-4}$    D) $\dfrac{x-1}{x-5}$

15.

16. If $b = \underline{\ ?\ }$, then $x^2 + bx + 4 = 0$ has two equal solutions.

A) –6    B) –2    C) 0    D) 4

16.

17. If $\dfrac{x}{y} = \dfrac{2}{9}$ and $\dfrac{z}{y} = \dfrac{4}{5}$, then $\dfrac{x}{z} =$

A) $\dfrac{5}{18}$    B) $\dfrac{8}{45}$    C) $\dfrac{6}{14}$    D) $\dfrac{7}{13}$

17.

18. Which of the following is a factor of $(x^3 - 4)^2 + 8x^3 - 25$?

A) $x^3 + 5$    B) $x^3 + 4$    C) $x^3 + 3$    D) $x^3 + 2$

18.

19. If $x$, $y$, and $z$ are integers and the average of $x$ and $y$ is $z$, then $x - y$ *cannot* equal

A) 0    B) 1    C) $y$    D) $z$

19.

20. If 24 workers can build a house in 10 hours, and all workers work at the same rate, then 40 workers could build the house in $\underline{\ ?\ }$ hours.

A) $\dfrac{50}{3}$    B) 6    C) 14    D) $\dfrac{64}{10}$

20.

21. If $x^2 + 8x + 16 = 5^{12}$ and $x > 0$, then $x =$

A) $5^6 - 4$    B) $5^6$    C) $5^6 + 4$    D) $5^6 + 8$

21.

22. If $x$ is a prime number greater than 3, how many different positive divisors does $6x$ have?

A) 2    B) 4    C) 6    D) 8

22.

Go on to the next page ))))➤ **A**

23. On a flat surface, my duck Fred walks due north at 1 km per hour for 1 minute, then walks due east at 0.75 km per hour for 1 minute. Fred is then how many km from his starting point?

    A) $\frac{1}{100}$     B) $\frac{1}{70}$     C) $\frac{1}{48}$     D) $\frac{1}{20}$

**23.**

24. $\sqrt{100^{100}} =$

    A) $10^{10}$     B) $10^{50}$     C) $100^{10}$     D) $100^{50}$

**24.**

25. In the complete expansion of $(x + 1)^4$, what is the sum of the coefficients of the odd powers of $x$?

    A) 4          B) 6          C) 8          D) 10

**25.**

26. If $x + 4 = y + 5$, then $x^2 + 8x =$

    A) $y^2 + 10y + 9$     B) $y^2 + 9y + 10$     C) $y^2 + 10y$     D) $y^2 + 9y$

**26.**

27. What is the greatest positive integer $x$ such that $2^x$ is a divisor of $12^{1200}$?

    A) 600          B) 1200          C) 2400          D) 3600

**27.**

28. One hundred grandparents were shopping. If $\frac{2}{3}$ of the grandfathers and $\frac{1}{2}$ of the grandmothers bought coats, and the number of grandfathers who bought coats was 10 more than $\frac{3}{2}$ the number of grandmothers who bought coats, how many grandfathers bought coats?

    A) 35     B) 40     C) 55     D) 60

**28.**

29. If $r$ is the sum of all even integers between 1 and 2011, and $s$ is the sum of all odd integers between 0 and 2010, then $r - s =$

    A) 1005          B) 1006          C) 2010          D) 2011

**29.**

30. If $x \neq 0$, $y \neq 0$, $x \neq y$, and $\frac{y}{x} - \frac{x}{y} + \frac{1}{x} - \frac{1}{y} = 0$, what is the value of $x + y$?

    A) –2          B) –1          C) 0          D) 1

**30.**

*The end of the contest* ✒ **A**

**Visit our Web site at http://www.mathleague.com**

Solutions on Page 129 • Answers on Page 151

# 2010-2011 Annual Algebra Course 1 Contest

*Spring, 2011*

## Instructions

- **Time**  Do *not* open this booklet until you are told by your teacher to begin. You will have only *30 minutes* working time for this contest. You might be *unable* to finish all 30 questions in the time allowed.

- **Scores**  Please remember that *this is a contest, and not a test*—there is no "passing" or "failing" score. Few students score as high as 24 points (80% correct). Students with half that, 12 points, *should be commended!*

- **Format and Point Value**  This is a multiple-choice contest. Each answer is an A, B, C, or D. Write each answer in the *Answer Column* to the right of each question. A correct answer is worth 1 point. Unanswered questions receive no credit. You **may** use a calculator.

1. If $xy = 2011^2$, then $(-x)(-y) =$

    A) $-2011^2$     B) $2011^{-2}$     C) $-2011^{-2}$     D) $2011^2$

2. There are _?_ days in $w$ weeks.

    A) $\dfrac{w}{7}$     B) $w + 7$     C) $7w$     D) $w^7$

3. Edna rides $d$ km on her scooter. If $(x + 2)(x - d) = x^2 - 4x - 12$ for all real numbers $x$, then $d =$

    A) -8     B) -6     C) 6     D) 8

4. $\sqrt{4x} + \sqrt{9x} + \sqrt{25x} =$

    A) $10\sqrt{x}$     B) $10 + \sqrt{x}$     C) $10 + 3\sqrt{x}$     D) $30\sqrt{x}$

5. $x - 2x + 3x - 4x + 5x - 6x + 7x - 8x + 9x - 10x =$

    A) $-10x$     B) $-5x$     C) $-2x$     D) 0

6. The sum of five consecutive integers is 165. The largest of these five integers is

    A) 31     B) 33     C) 35     D) 37

7. $\dfrac{x}{z} \div \dfrac{z}{x} =$

    A) 1     B) $\dfrac{x^2}{z^2}$     C) $\dfrac{z^2}{x^2}$     D) $x^2z^2$

8. Lois and Clark are flying $h$ m above the ground. If $(h + 2)^2 = 28^2$, then $h =$

    A) 26     B) 52     C) $26^2$     D) $52^2$

9. $(x + 5)^2 - (x - 5)^2 =$

    A) 0     B) 50     C) $10x$     D) $20x$

10. If $p$ is a prime number between 1000 and 2000, then _?_ could *not* be a prime number.

    A) $p - 10$    B) $p + 300$    C) $p + 456$    D) $p + 567$

11. Twice my age plus three times my sister's age is 86. Three times my age plus four times my sister's age is 120. How old am I?

    A) 15     B) 16     C) 17     D) 18

*Go on to the next page* )))➡ **A**

| | Answers |
|---|---|
| 12. If $n^2 + 5n = 24$ and $n^2 - 4n = -3$, then $9n =$ <br><br> A) -9    B) 9    C) 27    D) 81 | 12. |
| 13. If $a < 0 < b$, then which of the following *cannot* be true? <br><br> A) $\dfrac{1}{a} > \dfrac{1}{b}$    B) $a^2 > b^2$    C) $\dfrac{a}{b} > \dfrac{b}{a}$    D) $\dfrac{a^2}{b^2} > \dfrac{b^2}{a^2}$ | 13. |
| 14. Nervous Ned must find a polynomial that is divisible by $x - 3$. Of the following polynomials, which could Ned choose? <br><br> A) $x^2 + 9$        B) $x^2 + 8x + 15$ <br> C) $x^3 - 4x^2 + 4x - 3$    D) $x^3 - 2x^2 + 3x - 4$ | 14. |
| 15. The sum of the solutions of $x^2 + 2011x + 2010 = 0$ is <br><br> A) -2011    B) -2010    C) 2010    D) 2011 | 15. |
| 16. The line __?__ is perpendicular to the line $y = x$. <br><br> A) $3x + 4y = 5$    B) $5x - 5y = 13$    C) $6x + 6y = 19$    D) $8x - 7y = 31$ | 16. |
| 17. If $y = x - 5$, then $x^2 - 10x + 20 =$ <br><br> A) $y^2$        B) $y^2 - 5$        C) $y^2 - 10y + 25$    D) $y^2 + 10y + 15$ | 17. |
| 18. If $a + b = 7$ and $a^2 + b^2 = 49$, then $ab =$ <br><br> A) 42    B) 36    C) 7    D) 0 | 18. |
| 19. If $x$ is $y$ less than $2z$, what is the value of $z$ in terms of $x$ and $y$? <br><br> A) $2x - y$    B) $\dfrac{x+y}{2}$    C) $\dfrac{y-x}{2}$    D) $\dfrac{x}{2} + y$ | 19. |
| 20. Rob Esch searches a trash can for his cell phone. When he starts searching, there are 5 apple cores for every 3 bottles. If 3 more bottles are thrown in, there will be 3 apple cores for every 2 bottles. How many apple cores are there when he starts? <br><br> A) 15    B) 25    C) 35    D) 45 | 20. |
| 21. If $x > 0$, what percent of $0.2x$ is $2x$? <br><br> A) 1000%    B) 900%    C) 100%    D) 10% | 21. |
| 22. $|2x| + |-3x| =$ <br><br> A) $|-x|$    B) $-|x|$    C) $-|5x|$    D) $5|x|$ | 22. |

*Go on to the next page* ))))➡ **A**

67

23. Maria loves finding the roots of equations. If she finds all the integral roots of the equation $(x^2 - 1)^1 \times (x^2 - 2)^2 \times (x^2 - 3)^3 \times \ldots \times (x^2 - 20)^{20} = 0$, how many different integers will she find?

    A) 4    B) 8    C) 20    D) 40

23.

24. The product of all values of $x$ that satisfy $2011^{x^2+10x+21} = 1$ is

    A) -21    B) -10    C) 10    D) 21

24.

25. If $n$ is the smallest positive integer such that $99n$ is the cube of an integer, and $d$ is the sum of the digits of $n$, then $d$ is

    A) 27    B) 18    C) 12    D) 9

25.

26. The area of my rectangle is 480. If my rectangle's length is 14 greater than its width, then its perimeter is

    A) 88    B) 92    C) 116    D) 172

26.

27. If $18x + 27y + 38 = 74$, then $4x + 6y - 8 =$

    A) 0    B) 12    C) 24    D) 36

27.

28. Eunice always juggles as many things as she can handle. The number of things she juggles is the same as the value of $|2x-9|-|2x+9|$, where $x$ is some real number. Of the following, which can be the number of things she juggles?

    A) 15    B) 19    C) 20    D) 22

28.

29. What is the ones digit in the decimal representation of $s$ if $r = 123^{124}$ and $s = r^{456}$?

    A) 1    B) 3    C) 7    D) 9

29.

30. The least common multiple of all integers from 1 through 30 is divided by the product of all prime numbers between 1 and 30. The resulting quotient is

    A) 1    B) 2    C) 12    D) 360

30.

*The end of the contest* ✍ **A**

# Detailed Solutions

## 2006-2007 through 2010-2011

# 7th Grade Solutions

## 2006-2007 through 2010-2011

# Information & Solutions

*Tuesday, February 20 or 27, 2007*

**7**

## Contest Information

- **Solutions** Turn the page for detailed contest solutions (written in the question boxes) and letter answers (written in the *Answers* column to the right of each question).

- **Scores** Please remember that *this is a contest, not a test*—and there is no "passing" or "failing" score. Few students score as high as 30 points (75% correct). Students with half that, 15 points, *deserve commendation!*

- **Answers & Rating Scale** Turn to page 138 for the letter answers to each question and the rating scale for this contest.

| | |
|---|---|
| 1. $99+199+999+1999 = (100+200+1000+2000)-4 = 3300-4.$ <br> A) 14    B) 9    C) 4    D) 1 | 1. <br> C |
| 2. 0.2727's rightmost digit is 7; round UP to 0.273. <br> A) 0.272   B) 0.273   C) 0.2730   D) 0.2737 | 2. <br> B |
| 3. $4 \times 44 = 4 \times 2 \times 22 = 8 \times 22.$ <br> A) $16 \times 4$    B) $12 \times 33$    C) $11 \times 8$    D) $8 \times 22$ | 3. <br> D |
| 4. 80 hundredths $= 0.80 = 8/10 = 4/5.$ <br> A) 4    B) 8    C) 16    D) 20 | 4. <br> A |
| 5. Do $\times$ and $\div$ in order of appearance, *left to right.* <br> A) 4    B) 8    C) 16    D) 64 | 5. <br> A |
| 6. I can buy 12 balloons for \$10 plus 3 more for \$3 (or \$2.50). So, I can buy 15 balloons for \$13. <br> A) 16 balloons   B) 15 balloons <br> C) 14 balloons   D) 13 balloons | 6. <br> B |
| 7. Factoring out a 10, $(5 \times 10)+(5 \times 20)+(5 \times 30) = 10 \times (5+10+15).$ <br> A) $(5+50+100)$   B) $(5+20+30)$    C) $(5+10+20)$    D) $(5+10+15)$ | 7. <br> D |
| 8. Half of $16 \times 3 \ \ell =$ half of $48 \ \ell = 24 \ \ell = 12 \times 2 \ \ell.$ <br> A) 8    B) 12    C) 16    D) 24 | 8. <br> B |
| 9. Multiplying by $\frac{6}{4}$ is the same as dividing by its reciprocal, $\frac{4}{6} = \frac{2}{3}.$ <br> A) $\frac{2}{3}$    B) $\frac{3}{2}$    C) $\frac{3}{4}$    D) $\frac{4}{3}$ | 9. <br> A |
| 10. The factors of 60 which are multiples of 6 are 6, 12, 30, and 60. <br> A) 2    B) 3    C) 4    D) 5 | 10. <br> C |
| 11. $(3^2+4^2+5^2) \times (25-16-9) = (3^2+4^2+5^2) \times 0 = 0.$ <br> A) $9+16+25$   B) $6+8+10$   C) $3+4+5$   D) 0 | 11. <br> D |
| 12. Since $563 = (80 \times 7) + 3$, it's 80 weeks plus 3 days before Tuesday. Counting backwards, 3 days before Tuesday is Saturday. <br> A) Saturday    B) Friday <br> C) Wednesday    D) Tuesday | 12. <br> A |
| 13. $(\frac{1}{2} + \frac{1}{4}) - (\frac{1}{2} \times \frac{1}{4}) = \frac{3}{4} - \frac{1}{8} = \frac{6}{8} - \frac{1}{8} = \frac{5}{8}.$ <br> A) 0    B) $\frac{3}{8}$    C) $\frac{5}{8}$    D) $\frac{3}{4}$ | 13. <br> C |
| 14. Half of $20 = 10$. Now, $20\% =$ one-fifth, and $10 =$ one-fifth of 50. <br> A) 40    B) 50    C) 80    D) 100 | 14. <br> B |

*Go on to the next page* ⇒ **7**

| | | |
|---|---|---|
| 15. | A pentagon has 5 sides.<br>A) rhombus    B) octagon     C) hexagon     D) pentagon | 15.<br>D |
| 16. | Since 19 is prime, its largest prime factor is the number itself, 19.<br>A) 7          B) $14 = 2 \times 7$   C) 19         D) $20 = 2^2 \times 5$ | 16.<br>C |
| 17. | All sides have the same length, so 100 is a multiple of this length.<br>Since 100 is divisible by 4, but not 3, the side-lengths could be 25.<br>A) 20          B) 25          C) 33          D) 50 | 17.<br>B |
| 18. | (7 secs)/(1 week) = (7 secs)/$(7 \times 24 \times 60 \times 60)$ secs $= 1/86\,400$.<br>A) $\frac{1}{24}$   B) $\frac{1}{1440}$   C) $\frac{1}{3600}$   D) $\frac{1}{86\,400}$ | 18.<br>D |
| 19. | Both March and May have 31 days,<br>April has 30 days, and June 1 is 1<br>more day. Adding these, it takes<br>$31 + 31 + 30 + 1 = 93$ giant apples.<br>A) 93   B) 92   C) 91   D) 90 | 19.<br>A |
| 20. | $36 = \sqrt{36 \times 36} = \sqrt{1 \times 2 \times 3 \times 216}$.<br>A) 1296    B) 216    C) 36    D) 6 | 20.<br>B |
| 21. | Average of $\frac{5}{4}, \frac{7}{4}, \frac{9}{4}, \frac{11}{4} = $ sum $\div 4 = \frac{32}{4} \div 4 = 2 = $ average of 1 & 3.<br>A) 3          B) 4          C) 9          D) 15 | 21.<br>A |
| 22. | (# dimes) $= 2.5 \times$ (# quarters), for ANY amount of money.<br>A) 0.25        B) 1.5         C) 2.5         D) 10 | 22.<br>C |
| 23. | Choices A, B, C are always even; but $6 \div 2 = 3$, which is odd.<br>A) $(number)^2$   B) $\sqrt{number}$   C) $2 \times number$   D) $number \div 2$ | 23.<br>D |
| 24. | Each of the 4 small squares has 4 right angles.<br>Now, add in the 4 right angles from the diamond,<br>and the total is $4 \times 4 + 4 = 20$ right angles.<br>A) 20      B) 16      C) 12      D) 8 | 24.<br>A |
| 25. | $(2^3 \times 3^3 \times 5^3) \div (3 \times 4 \times 5 \times 6) = (3 \times 5^2)$.<br>A) 100    B) 75    C) 50    D) 25 | 25.<br>B |
| 26. | Since 40% = 30 days, multiply both<br>sides by 2.5 to get 100% = 75 days.<br>A) 12    B) 42    C) 50    D) 75 | 26.<br>D |
| 27. | $\left(\frac{2}{3} \times \frac{3}{2}\right) \times \left(\frac{2}{3} \times \frac{3}{2}\right) \times \left(\frac{2}{3} \times \frac{3}{2}\right) \times \frac{2}{3} = 1 \times 1 \times 1 \times \frac{2}{3}$.<br>A) $\frac{1}{3}$    B) $\frac{2}{3}$    C) 1    D) $\frac{3}{2}$ | 27.<br>B |
| 28. | $5^4 \div 10^4 = 1^4 \div 2^4 = 1 \div 16 = 0.0625 = 0.0625 \times 100\% = 6.25\%$.<br>A) 0.5         B) 6.25        C) 25        D) 50 | 28.<br>B |

29. The area of a semicircle $= \pi r^2/2 = 64\pi/2 = 32\pi$.
    A) $32\pi$      B) $64\pi$      C) $128\pi$      D) $256\pi$

29. A

30. $r(r(2) \div r(3) \div r(4)) = r(\frac{1}{2} \div \frac{1}{3} \div \frac{1}{4}) = r(\frac{3}{2} \div \frac{1}{4}) = r(\frac{12}{2}) = \frac{1}{6}$.
    A) $\frac{2}{3}$    B) $\frac{3}{8}$    C) $\frac{1}{6}$    D) $\frac{1}{24}$

30. C

31. Cupid shot his arrow 10 000 times. His success rate was $\sqrt{1\%} = \sqrt{1/100} = 1/10$. He succeeded $(1/10) \times 10\,000 = 1000$ times.
    A) 1      B) 10      C) 100      D) 1000

31. D

32. Every number is a factor of itself.
    A) 5      B) 1000      C) 5000      D) 10 000

32. D

33. There were 2 horses, $2 \times 4 = 8$ cows, $8 \times 8 = 64$ pigs, and $16 \times 64 = 1024$ chickens. Altogether, the total number of animals was $2 + 8 + 64 + 1024 = 1098$.
    A) 30      B) 512      C) 1024      D) 1098

33. D

34. With 9 sides, there can be at most 4 such pairs. For an example, we can remove a STOP sign's corner.
    A) 2      B) 3      C) 4      D) 8

34. C

35. Since $4 \times 5 \times 6 \times 7 \times 8$ is not divisible by 9, the correct answer must be choice B.
    A) 5      B) 4      C) 3      D) 2

35. B

36. Cory's test total is $15 \times 85 = 1275$. His known scores total 855. The other 5 scores must total $1275 - 855 = 420$, an average of 84.
    A) 82      B) 84      C) 85      D) 86

36. B

37. Switch the divisor and quotient to get $\underline{\ ?\ } = \frac{2007}{2006} \div \frac{2006}{2007} = \frac{2007^2}{2006^2}$.
    A) $\frac{2007^2}{2006^2}$    B) $\frac{2006^2}{2007^2}$    C) $\frac{2 \times 2006^2}{2007^2}$    D) 1

37. A

38. The lcd of $\frac{1}{2}, \frac{2}{3}, \frac{3}{4}, \frac{4}{5}, \frac{5}{6}, \frac{6}{7}, \frac{7}{8}, \frac{8}{9}$, and $\frac{9}{10}$ is $5 \times 7 \times 8 \times 9 = 2520$.
    A) 10      B) 1260      C) 2520      D) 3 628 800

38. C

39. Except for 1, **every** positive integer is a multiple of a prime.
    A) 1      B) 499      C) 998      D) 999

39. C

40. There is no carryover from one column to another, so $R + S + T + U = (R+U) + (S+T) = T + T = 2 \times T$.
    A) $2 \times (R+S)$   B) $2 \times (S+U)$   C) $2 \times (T+U)$   D) $2 \times T$

    $$\begin{array}{r} RSTU \\ +\,UTSR \\ \hline TTTT \end{array}$$

40. D

*The end of the contest* ☞ **7**

**Visit our Web site at http://www.mathleague.com**

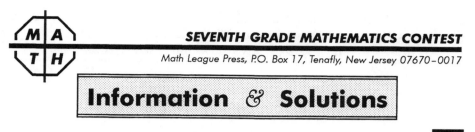

# Information & Solutions

*Tuesday, February 19 or 26, 2008*

## Contest Information

**7**

- **Solutions** Turn the page for detailed contest solutions (written in the question boxes) and letter answers (written in the *Answers* column to the right of each question).

- **Scores** Please remember that *this is a contest, not a test*—and there is no "passing" or "failing" score. Few students score as high as 30 points (75% correct). Students with half that, 15 points, *deserve commendation!*

- **Answers & Rating Scale** Turn to page 139 for the letter answers to each question and the rating scale for this contest.

1. $(10 \times 2008) + (1 \times 2008) = (10+1) \times 2008 = 11 \times 2008$.

   A)  4026     B)  $10 \times 2009$     C)  $10 \times 4016$     D)  $11 \times 2008$

   **1.** D

2. $30¢ = 10¢ + 10¢ + 10¢$.

   $40¢ = 5¢ + 10¢ + 25¢$.

   $60¢ = 10¢ + 25¢ + 25¢$.

   No 3 coins total $50¢$ in value.

   A) 30   B) 40   C) 50   D) 60

   **2.** C

3. Since 8 is already a multiple of both 2 and 4, the lcm is 8.

   A) 2     B) 8     C) 14     D) 64

   **3.** B

4. $0.10 + 0.10 = 0.20 = 0.200$.

   A)  0.200     B)  0.020     C)  0.110     D)  0.100

   **4.** A

5. Since $5 \times 7 = 35$, the numbers are 5 and 7. Their sum is $5+7 = 12$.

   A)  8     B)  12     C)  18     D)  36

   **5.** B

6. Only in D do we add the 1. Multiplying by 1 has no effect.

   A) $0.1 \times 1.1 \times 1$   B) $0.1 + 1.1 \times 1$   C) $1.1 + 0.1 \times 1$   D) $0.1 + 1.1 + 1$

   **6.** D

7. $20$ cm $\div 1\frac{1}{4}$ cm $= 16$.

   A)  15     B)  16     C)  18     D)  25

   **7.** B

8. The average of my 3 tests is 90, so their sum is $3 \times 90 = 270$.

   A)  270     B)  180     C)  90     D)  30

   **8.** A

9. The length of each side of this pool is 6 m, the square root of 36 m$^2$. The perimeter of this pool is $4 \times 6$ m $= 24$ m.

   A) 18 m     B) 24 m     C) 36 m     D) 81 m

   **9.** B

10. If Dan lives on a line between Al and Bob, Dan is 4 km from Bob and 1 km from Al.

    A) 1 km     B) 2 km     C) 3 km     D) 4 km

    **10.** A

11. The ratio is $(24/1):(1/24) = (24/1) \times (24/1) = 576/1 = 576:1$.

    A)  1:24     B)  24:1     C)  48:1     D)  576:1

    **11.** D

12. $58 = 2 \times 29$, and 29 is prime, so 58 has the largest prime factor.

    A) $49 = 7 \times 7$     B) $51 = 3 \times 17$     C) $58 = 2 \times 29$     D) $65 = 5 \times 13$

    **12.** C

13. $\frac{1}{2}$ of $\frac{1}{3} = \frac{1}{2} \times \frac{1}{3} = \frac{1}{3} \times \frac{1}{2} = \frac{1}{3}$ of $\frac{1}{2}$.     A) $\frac{1}{6}$ B) $\frac{1}{4}$ C) $\frac{1}{3}$ D) $\frac{1}{2}$

    **13.** D

14. The product of 4 and $4^2$ is $4 \times 16 = 64$.

    A)  2     B)  4     C)  6     D)  8

    **14.** B

15. The ten integers between $1\frac{1}{9}$ and $11\frac{1}{9}$ are 2, 3, . . . , 10, 11.

    A)  8     B)  9     C)  10     D)  90

    **15.** C

| | |
|---|---|
| 16. $\sqrt{9} + \sqrt{16} = 3+4 = 7 = 5+2 = \sqrt{25} + \sqrt{4}$.<br>A) $\sqrt{0}$  B) $\sqrt{2}$  C) $\sqrt{4}$  D) $\sqrt{49}$ | 16.<br>C |
| 17. Any positive number bigger than 1 will exceed its reciprocal.<br>A) 1  B) 2  C) 3  D) 4 | 17.<br>A |
| 18. 4:18 P.M. − 3:26 P.M. = 52 minutes,<br>so the correct answer is 52 minutes<br>before 3:26 P.M. That's 2:34 P.M.<br>A) 2:18  B) 2:32<br>C) 2:34  D) 2:44 | 18.<br><br>C |
| 19. The sum of the lengths of any<br>2 sides must be > the 3rd side.<br>A) 4, 5, 6  B) 3, 4, 5<br>C) 2, 3, 4  D) 1, 2, 3 | 19.<br><br>D |
| 20. $\frac{31}{16} + \frac{32}{16} + \frac{33}{16} = \frac{96}{16} = \frac{48}{8}$.  A) 48 B) 36 C) 16 D) 12 | 20. A |
| 21. $\frac{36}{54} = \frac{2}{3}$ or $\frac{4}{6}$ or $\frac{6}{9}$ or $\frac{8}{12}$ or $\frac{10}{15}$ or $\frac{12}{18}$ or $\frac{14}{21}$ or $\frac{16}{24}$.<br>A) 6  B) 8  C) 12  D) 15 | 21.<br>B |
| 22. The sum of an odd and an even number is always odd, never even.<br>A) 1  B) prime  C) odd  D) even | 22.<br>D |
| 23. Use the fraction whose numerator is 12 *before* it's increased by 12.<br>A) $\frac{3}{41}$  B) $\frac{6}{41}$  C) $\frac{12}{41}$  D) $\frac{24}{41}$ | 23.<br>C |
| 24. A triangle whose angles measure 70°, 70°, and 40° is isosceles.<br>A) 85°, 50°  B) 80°, 55°  C) 75°, 35°  D) 70°, 40° | 24.<br>D |
| 25. 10% of 10% = 0.10 × 0.10 = 0.01 = 1 × 0.01 = 100% of 1/100.<br>A) 1  B) $\frac{1}{10}$  C) $\frac{1}{100}$  D) $\frac{1}{1000}$ | 25.<br>C |
| 26. $10^{2008}$ has 2009 digits: a 1 followed by 2008 0s.<br>A) 2009  B) 2008  C) 2007  D) 20 080 | 26. A |
| 27. Bob must be the average height, so his<br>height is 150 cm. Carl is 2 cm shorter<br>than Bob, so Carl's height is 148 cm.<br>A) 152  B) 148  C) 147  D) 146 | 27.<br><br>B |
| 28. The numerator is half the denominator.<br>A) 12  B) 14  C) 26  D) 27 | 28.<br>C |
| 29. $45^2 = 2025$ and $44^2 = 1936$; their difference is $2025 - 1936 = 89$.<br>A) 88  B) 89  C) 90  D) 91 | 29.<br>B |

*Go on to the next page* ▮▮▮➡ **7**

| | |
|---|---|
| 30. A circle can pass through 1, 2, or 4 of a square's vertices, not just 3.<br>A) 4 　　B) 3 　　C) 2 　　D) 1 | 30.<br>B |
| 31. 7/8 is nearly 1, whereas 8/17 < 1/2 and 6/11 > 1/2 are clearly smaller.<br>A) $\frac{7}{8}<\frac{8}{17}<\frac{6}{11}$ 　B) $\frac{6}{11}<\frac{8}{17}<\frac{7}{8}$ 　C) $\frac{6}{11}<\frac{7}{8}<\frac{8}{17}$ 　D) $\frac{8}{17}<\frac{6}{11}<\frac{7}{8}$ | 31.<br>D |
| 32. Face of large cube: area = 9600/6 = 1600; edge = $\sqrt{1600}$ = 40.<br>Face of small cube: area = 96/6 = 16; edge = 4; $40^3 \div 4^3 = 1000$.<br>A) 10 　B) 100 　C) 1000 　D) 10 000 | 32.<br>C |
| 33. To buy a $45 gift, I paid 25% or $11.25, leaving $33.75. Each of my 5 payments is $33.75÷5.<br>A) $6.75 　B) $7.25 　C) $9.00 　D) $11.25 | 33.<br>A |
| 34. If $N = 2$, only $N+7 = 9$ is not a prime number.<br>A) $N+3$ 　　B) $N+5$<br>C) $N+7$ 　　D) $N+9$ | 34.<br>C |
| 35. $(2 \times \frac{1}{3}) \times (4 \times \frac{1}{5}) \times \ldots \times (48 \times \frac{1}{49}) < 1$ since each subproduct < 1.<br>A) < 1 　B) > 1 　C) = 1 　D) = 0 | 35.<br>A |
| 36. Since 120 kids have a cat, 300 have a dog, and 65 have both, 120−65 = 55 have only a cat and 300−65 = 235 have only a dog. Hence, 400−(55+235+65) = 45 have neither pet.<br>A) 45 　B) 98 　C) 129 　D) 335 | 36.<br>A |
| 37. $(2 \times 2 \times 2)^8 \times (2 \times 2)^4 \times 2^2 = 2^{8+8+8+4+4+2} = 2^{34}$.<br>A) $2^{64}$ 　B) $2^{34}$ 　C) $2^{26}$ 　D) $2^{14}$ | 37.<br>B |
| 38. $\sqrt{\sqrt{\sqrt{100 \times 100 \times 100 \times 100}}} = \sqrt{\sqrt{100 \times 100}} = \sqrt{100} = 10$.<br>A) 1 　B) $\sqrt{10}$ 　C) 10 　D) 100 | 38.<br>C |
| 39. Every 360°, each side returns to its original position. Next, 1575° is 135° more than 4×360° = 1440°. The extra 135° is 3/8 of 360°, so each side moves 3 positions clockwise, and side 8 ends where side 3 began.<br>A) side 2 　B) side 4 　C) side 6 　D) side 8 | 39.<br>D |
| 40. I swam 58 laps on May 16, the middle day of May. I swam 1 less lap each day before this, so I swam 58−15 = 43 laps on May 1.<br>A) 42 　B) 43 　C) 44 　D) 45 | 40.<br>B |

*The end of the contest* ✍ **7**

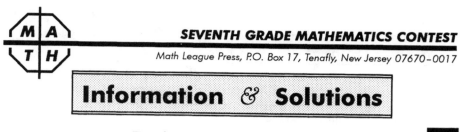

## Information & Solutions

*Tuesday, February 17 or 24, 2009*

### Contest Information

**7**

- **Solutions** Turn the page for detailed contest solutions (written in the question boxes) and letter answers (written in the *Answers* column to the right of each question).

- **Scores** Please remember that *this is a contest, not a test*—and there is no "passing" or "failing" score. Few students score as high as 30 points (75% correct). Students with half that, 15 points, *deserve commendation!*

- **Answers & Rating Scale** Turn to page 140 for the letter answers to each question and the rating scale for this contest.

| | |
|---|---|
| 1. $(96+4)+(97+3)+(98+2)+(99+1) = 100+100+100+100 = 4\times100.$ <br> A) 50      B) 95      C) 96      D) 100 | 1. <br> D |
| 2. Since $6\times33 = 198$, and $200-198 = 2$, 2 dimes are left over. <br> A) 1      B) 2      C) 3      D) 4 | 2. <br> B |
| 3. The hour hand moves 360° every 12 hours. It moves $360°\div12 = 30°$ each hour. <br> A) 1    B) 2    C) 3    D) 6 | 3. <br> B |
| 4. The average number of days is $(365+365+365+366) \div 4 = 365.25.$ <br> A) 365.00      B) 365.10 <br> C) 365.25      D) 365.75 | 4. <br> C |
| 5. $\frac{1}{4}$ of 300 is 75, and $\frac{1}{3}$ of 75 is 25. <br> A) 25    B) 75    C) $\frac{1}{400}$    D) $\frac{1}{3600}$ | 5. <br> A |
| 6. $4.5 - 1.5 = 3$, which is 6 halves. <br> A) 1      B) 3      C) 6      D) 9 | 6. <br> C |
| 7. $2412 \div 4 = 603$ and $2412 \div 6 = 402.$ <br> A) 1246      B) 2412      C) 4664      D) 6424 | 7. <br> B |
| 8. $(1\times10\times100)\times1000 = 1000\times1000 = 1\,000\,000.$ <br> A) 1111      B) 11\,110      C) 1\,000\,000      D) 1\,111\,000\,000 | 8. <br> C |
| 9. $4\times(4+5) + 5\times(4+5) = 4\times9 + 5\times9 = 9\times9 = 9^2.$ <br> A) $9^2$      B) 99      C) $9 + 9^2$      D) $9^3$ | 9. <br> A |
| 10. $15 = 1+2+3+4+5$ can be the sum of 5 consecutive positive integers. <br> A) 2    B) 3    C) 4    D) 5 | 10. <br> D |
| 11. Since 25% of $40 = 0.25\times\$40 = \$10$, Chef Smile's salary is $\$40 + \$10 = \$50.$ <br> A) \$15    B) \$30    C) \$50    D) \$65 | 11. <br> C |
| 12. Since $2\div2 = 1$, choices B, C, and D are all possible. <br> A) even      B) odd <br> C) whole      D) 1 | 12. <br> A |
| 13. $18^2\times9 = 9^2\times2^2\times9 = 9^2 \times36 = 9^2\times6^2.$ <br> A) 18    B) $2^2$    C) $3^2$    D) $6^2$ | 13. <br> D |
| 14. It's true that $\frac{1}{100}$ of 8 is $\frac{8}{100}$. <br> A) $\frac{1}{100}$      B) $\frac{8}{100}$      C) $\frac{64}{100}$      D) $\frac{8}{10}$ | 14. <br> A |

15. Each $\angle$ of an equilateral $\triangle$ is 60°. Since $m\angle BCD = m\angle BCA + m\angle ACD$, $m\angle BCD = 60 + 60 = 120$.

A) 60     B) 120     C) 150     D) 180

15. B

16. From 3:15 P.M. to 4:30 P.M. is 75 mins. Walking uphill took twice as long as walking downhill. We walked uphill 50 mins. and downhill 25 mins. We began to walk downhill at 3:15 + 50 mins. = 4:05 P.M.

A) 3:40 P.M.     B) 3:55 P.M.
C) 4:00 P.M.     D) 4:05 P.M.

16. D

17. $1\,km \div 1\,cm = 1000\,m \div 0.01\,m = 100\,000$.

A) 100 000     B) 10 000
C) 1000     D) 100

17. A

18. $3000 + 0.030 = 3000.030$.

A) 33 000     B) 3000.03     C) 3000.003     D) 0.330

18. B

19. $(1\times 24)\times(2\times 12)\times(3\times 8)\times(4\times 6) = 24\times 24\times 24\times 24 = 24^4$.

A) 24     B) $24^2$     C) $24^3$     D) $24^4$

19. D

20. $\sqrt{16}\times(\sqrt{8}\times\sqrt{2})\times\sqrt{4} = (\sqrt{16}\times\sqrt{16})\times 2 = 16\times 2$.

A) 2     B) 4     C) 8     D) 64

20. A

21. Here, do addition last: $12+[(72\div 6)\times 2] = 12+[12\times 2] = 12+24 = 36$.

A) 7     B) 18     C) 28     D) 36

21. D

22. $\frac{11}{10}+\frac{101}{100}+\frac{1001}{1000} = \frac{1100}{1000}+\frac{1010}{1000}+\frac{1001}{1000} = \frac{3111}{1000} = 3\frac{111}{1000}$.

A) $3\frac{1}{10}$     B) $3\frac{3}{10}$     C) $3\frac{1}{1000}$     D) $3\frac{111}{1000}$

22. D

23. $(2\times 12)\times(2\times 13)\times(2\times 14)\times(2\times 15) = (12\times 13\times 14\times 15)\times 2^4$.

A) 16     B) 12     C) 4     D) 2

23. A

24. 20% of $30 = 0.2\times 30 = 6 = 0.1\times 60 = 10\%$ of 60.

A) 1%     B) 10%     C) 40%     D) 100%

24. B

25. $30 = 90/3 = 60/2$. There are $90-60 = 30$ more thirds.

A) 5    B) 15    C) 20    D) 30

25. D

26. Area $= \pi\times r^2 = \pi$, so $r = 1$.
Diameter $= 2\times r = 2\times 1 = 2$.

A) 0.5    B) 1    C) 2    D) 4

26. C

27. $\frac{1}{2}+\frac{4}{2} = \frac{5}{2}$. Its reciprocal is $\frac{2}{5}$.

A) $\frac{2}{5}$    B) $\frac{2}{3}$    C) $\frac{3}{2}$    D) $\frac{5}{2}$

27. A

28. The polygons could be a square (4 sides) and a hexagon (6 sides).

A) a pentagon    B) a hexagon    C) an octagon    D) a decagon

28. B

*Go on to the next page* ⮞ **7**

| | |
|---|---|
| 29. $10^{12} = 2^{12} \times 5^{12} = (2\times2\times2\times2\times2\times2\times2\times2\times2\times2\times2\times2)\times$ $(5\times5\times5\times5\times5\times5\times5\times5\times5\times5\times5\times5)$, the product of 24 primes.<br>A) 20  B) 24  C) 48  D) 50 | 29.<br>B |
| 30. Al rode 48 km in 1.5 hrs. His rate was 48 km ÷ 1.5 hrs. = 32 km/hr.<br>A) 8 km/hr  B) 16 km/hr<br>C) 32 km/hr  D) 64 km/hr | 30.<br>C |
| 31. This can be achieved with any odd multiple of 3. Here, $99 = 31+33+35$.<br>A) 100  B) 99<br>C) 98  D) 97 | 31.<br>B |
| 32. 20% of 0.5% = $0.2\times0.5\%$ = 0.10% = 0.1%.<br>A) 0.1%  B) 1%  C) 10%  D) 10 | 32.<br>A |
| 33. $\frac{1}{2} \div \frac{2}{3} = \frac{1}{2} \times \frac{3}{2} = \frac{3}{4}$, and $\frac{2}{3} \times \frac{9}{8} = \frac{3}{4}$.<br>A) 2  B) $\frac{9}{8}$  C) $\frac{3}{4}$  D) $\frac{1}{2}$ | 33.<br>B |
| 34. The even #s are 2 times any of 1, 3, 5, 7, 3×5, 3×7, 5×7, or 3×5×7.<br>A) 1  B) 6  C) 7  D) 8 | 34.<br>D |
| 35. If I earn $100, you earn $0.50. I earn 200 times = $200\times100\%$ as much.<br>A) 2  B) 200  C) 2000  D) 20 000 | 35.<br>D |
| 36. Simplify to get $\frac{1}{2} \times \frac{2}{3} \times \frac{3}{4} \times\ldots\times \frac{98}{99} \times \frac{99}{100}$. A lot cancels!<br>A) $\frac{1}{100}$  B) $\frac{1}{99}$  C) $\frac{9}{10}$  D) $\frac{99}{100}$ | 36.<br>A |
| 37. (Al's age):(Ed's age) = 3:5 = 6:10 = 9:15 = ... = 27:45 = ....<br>A) 62  B) 72  C) 82  D) 92 | 37.<br>B |
| 38. $9^3 = 3\times3\times3\times3\times3\times3 = 27^2$. Side-lengths could be 27 and area $27^2$.<br>A) $7^3$  B) $8^3$  C) $9^3$  D) $10^3$ | 38.<br>C |
| 39. If 15 do both, 5 only sing, 5 only dance, and $50-15-5-5 = 25$ do neither.<br>A) 5  B) 10  C) 15  D) 25 | 39.<br>D |
| 40. They danced for 20 minutes on day 1 and (20+91) minutes on day 92 (91 days after day 1). The average is $(20+111)\div2$ minutes per day. They danced for 92 days × 65.5 minutes per day = 6026 minutes.<br>A) 1820  B) 5106  C) 6026  D) 6072 | 40.<br>C |

*The end of the contest* 7

**Visit our Web site at http://www.mathleague.com**

84

# Information & Solutions

## 2009-2010 Annual 7th Grade Contest

*Tuesday, February 16 or 23, 2010*

**7**

### Contest Information

- **Solutions**  Turn the page for detailed contest solutions (written in the question boxes) and letter answers (written in the *Answer Column* to the right of each question).

- **Scores**  Please remember that *this is a contest, and not a test*—there is no "passing" or "failing" score. Few students score as high as 30 points (75% correct); students with half that, 15 points, *deserve commendation!*

- **Answers and Rating Scales**  Turn to page 141 for the letter answers to each question and the rating scale for this contest.

1.  There are 25 squares in all, and 5 are shaded. So the percent shaded is $(5 \div 25) \times 100 = 20$.

    A) 5%  B) 20%  C) 25%  D) 50%

    1.
    B

2.  Work backwards: $4 \times 18 + 2 = 72 + 2 = 74$.

    A) 26          B) 56          C) 70          D) 74

    2.
    D

3.  2.345 is closer to 2.3 than it is to 2.4.

    A) 2.3  B) 2.34  C) 2.35  D) 2.5

    3.
    A

4.  The common elements are $b$ and $c$.

    A) { }   B) $\{b, c\}$  C) $\{a, d\}$  D) $\{a, b, c, d\}$

    4.
    B

5.  From 1 to 99 is 99 numbers. Subtract the first 20 numbers to get 79.

    A) 81     B) 80      C) 79      D) 78

    5.
    C

6.  $(8 \times 6 + 2) \div 2 = (48 + 2) \div 2 = 50 \div 2 = 25$.

    A) $4 \times 3 + 1 = 13$  B) $4 \times 6 + 1 = 25$  C) $4 \times 6 + 2 = 26$  D) $8 \times 6 + 1 = 49$

    6.
    B

7.  As shown below, only choice B is a prime number.

    A) $81 = 9 \times 9$      B) 83          C) $87 = 3 \times 29$      D) $99 = 9 \times 11$

    7.
    B

8.  1% of $2010 = 0.01 \times \$2010 = \$20.10$.

    A) $0.201        B) $2.01         C) $20.10         D) $201.00

    8.
    C

9.  $\sqrt{36} - \sqrt{25} = 6 - 5 = 1 = \sqrt{1}$.

    A) $\sqrt{1}$          B) $\sqrt{11}$          C) $\sqrt{14}$          D) $\sqrt{16}$

    9.
    A

10. There are 40 quarters in a roll. Each sister's share is $40 \div 8 = 5$, so three sisters take 15 coins, leaving 25 quarters in the roll. The value of these 25 quarters is $25 \times 25¢ = \$6.25$.

    A) $0.15  B) $1.25  C) $3.75  D) $6.25

    10.
    D

11. $2^4 + 2^4 = 16 + 16 = 32 = 2^5$.

    A) $2^5$   B) $4^8$   C) $4^4$   D) $2^8$

    11.
    A

12. A hexagon has 6 sides and an octagon has 8 sides. The ratio is $6:8 = 3:4$.

    A) 7:8   B) 5:6    C) 3:4    D) 1:2

    12.
    C

13. The average is $\left(\frac{1}{2} + \frac{1}{3}\right) \div 2 = \frac{5}{6} \times \frac{1}{2} = \frac{5}{12}$.

    A) $\frac{5}{6}$   B) $\frac{2}{5}$   C) $\frac{1}{6}$   D) $\frac{5}{12}$

    13.
    D

14. $(3 \times 2 \times 6 \times 2 \times 9 \times 2 \times 12 \times 2) \div (3 \times 6 \times 9 \times 12) = 2 \times 2 \times 2 \times 2 = 16$.

    A) 16          B) 8          C) 2          D) 0

    14.
    A

15. A circle's radius, $r$, divided by its circumference, $2\pi r$, has quotient $1/(2\pi)$.

    A) $2\pi$          B) $\pi$          C) $\frac{1}{\pi}$          D) $\frac{1}{2\pi}$

    15.
    D

Go on to the next page ))))➤ 7

Answers

16. Use estimation: $24 \div 0.5 = 48$.

   A) 12    B) 24    C) 36    D) 48

16.

D

17. Find two numbers whose product is 40 and whose sum is 14. The numbers are 4 and 10, so the length of the longest side is 10.

   A) 12 m   B) 10 m   C) 8 m    D) 5 m

17.

B

18. 300% of $30 = 3.00 \times 30 = 90$.

   A) 0.9    B) 9    C) 90    D) 9000

18.

C

19. Rewrite each fraction with a denominator of 3000.

   A) $\frac{1000}{3000}$   B) $\frac{1050}{3000}$   C) $\frac{990}{3000}$   D) $\frac{1001}{3000}$

19.

C

20. If 2 cm represent 6000 km, 1 cm represents 3000 km, and 0.1 cm represent 300 km.

   A) 0.1    B) 0.5    C) 10    D) 20

20.

A

21. All angles in an equilateral triangle are congruent, so $m\angle C = 60$. Thus, $m\angle A + m\angle B = 180 - 60 = 120$.

   A) 60    B) 90    C) 120    D) 180

21.

C

22. Since $300 \div 26$ has a quotient of 11 and a remainder of 14, the 300th letter written is the 14th one from the beginning of the alphabet, N.

   A) K         B) L         C) M         D) N

22.

D

23. The sum of Lana's first 4 grades is $4 \times 75 = 300$. The sum of her first 5 is $5 \times 80 = 400$. Her score on the 5th test is the difference: $400 - 300 = 100$.

   A) 100    B) 95    C) 85    D) 80

23.

A

24. $10$ m $+ 10$ cm $= 10$ m $+ 0.1$ m $= 10.1$ m.

   A) 11 m    B) 10.1 m   C) 10.01 m   D) 10.001 m

24.

B

25. $18{:}12 = 3{:}2 = (3 \times 12){:}(2 \times 12) = 36{:}24$.

   A) 2:3    B) 16:10    C) 12:18    D) 36:24

25.

D

26. $45^3 = (3^2 \times 5)^3 = 3^2 \times 3^2 \times 3^2 \times 5 \times 5 \times 5 = 3^6 \times 5^3$.

   A) $3^3 \times 5^3$   B) $4^3 \times 5^3$   C) $3^6 \times 5^3$   D) $3^8 \times 5^3$

26.

C

27. $(4 + 52)$ minutes after 10:56 AM is 11:52 AM.

   A) 10:00 AM    B) 11:52 AM    C) 11:56 AM    D) 12:02 PM

27.

B

28. The square of the reciprocal is $\frac{1}{9}$, so the square of the number is 9, the number is 3, and the number's cube is $3^3 = 3 \times 3 \times 3 = 27$.

   A) $\frac{1}{27}$        B) $\frac{1}{9}$        C) 9        D) 27

28.

D

29. It goes 240 000 m in 60 mins., 4000 m in 1 min., 2000 m in 30 seconds.

   A) 2         B) 8         C) 2000         D) 8000

29.

C

Go on to the next page ))))⮕ 7

87

30. If $A$, $B$, and $C$ are on a line, the distance from point $A$ to point $C$ can be 3 cm or 11 cm. Otherwise, the distance from point $A$ to point $C$ is between 3 cm and 11 cm.

    A) 2 cm          B) 3 cm          C) 6 cm          D) 10 cm

**30.**

**A**

31. Since S×RS has 3 digits and a ones digit of S, S is 5 or 6. If S is 5, R must be 2 or 3, since the product begins with 1. Only 3 works, since T > S. If S = 6 and R = 2, T = 5 < S, which is not allowed.

$$\begin{array}{r} RS \\ \times\ S \\ \hline 1TS \end{array}$$

    A) 1          B) 4          C) 5          D) 6

**31.**

**C**

32. Can the sides be 10, 10, and 20? No: the sum of the two smaller sides must be greater than the 3rd side. So the sides are 10, 20, and 20; the perimeter is 50.

    A) 60     B) 50     C) 40     D) 30

**32.**

**B**

33. $\dfrac{9}{2} - \dfrac{2}{9} = \dfrac{81}{18} - \dfrac{4}{18} = \dfrac{77}{18}$.

    A) $\dfrac{18}{77}$     B) $\dfrac{14}{18}$     C) $\dfrac{18}{14}$     D) $\dfrac{77}{18}$

**33.**

**D**

34. The numbers are the even sums 2+2, 2+4, 2+2, 2+4, 2+6, ... , 2+196. There are 98 such sums from 4 to 198.

    A) 101     B) 100     C) 99     D) 98

**34.**

**D**

35. In 1 hour, the wheel rolls 30 × 200$\pi$ m = 6000$\pi$ m. Its circumference is $(2 \times \pi \times 2)$ m = $4\pi$ m. It makes 6000$\pi$ m ÷ $4\pi$ m = 1500 full revolutions.

    A) 100          B) 200          C) 1500          D) 3000

**35.**

**C**

36. The product must be divisible by 210 because $2 \times 3 \times 5 \times 7 = 210$.

    A) 210     B) 260     C) 420     D) 520

**36.**

**A**

37. Of 180 paintings, 180 − 25 = 155 have blue and/or red borders. Since 110 + 90 = 200, 200 − 155 = 45 have borders with both colors.

    A) 25     B) 45     C) 55     D) 65

**37.**

**B**

38. The difference is (2+4+6+...+100) − (1+3+5+...+ 99) = (2−1) + (4−3) + (6−5) + ... + (100−99) = 1 + 1 + 1 + ... + 1 = 50.

    A) 100     B) 50     C) 25     D) 1

**38.**

**B**

39. Ones digits of powers of 2 cycle **2, 4, 8, 6**, 2, 4, 8, 6 ... ; 2009÷4 has R1.

    A) $2^{2009}$          B) $2^{2010}$          C) $2^{2011}$          D) $2^{2012}$

**39.**

**A**

40. Work backwards: Gwen had $\dfrac{3}{2} \times \$36 = \$54$ before buying clothes. Gwen had $\dfrac{7}{6} \times \$54 = \$63$ before buying food, so she spent $9 on food.

    A) $24          B) $18          C) $12          D) $9

**40.**

**D**

*The end of the contest* 🖉**7**

**Visit our Web site at http://www.mathleague.com**

# Information & Solutions

## 2010-2011 Annual 7th Grade Contest

*Tuesday, February 15 or 22, 2011*

**7**

### Contest Information

- **Solutions** Turn the page for detailed contest solutions (written in the question boxes) and letter answers (written in the *Answer Column* to the right of each question).

- **Scores** Please remember that *this is a contest, and not a test*—there is no "passing" or "failing" score. Few students score as high as 28 points (80% correct); students with half that, 14 points, *deserve commendation!*

- **Answers and Rating Scales** Turn to page 142 for the letter answers to each question and the rating scale for this contest.

1. $\dfrac{1}{2011} \times 2011^2 = \dfrac{1}{2011} \times 2011 \times 2011 = 1 \times 2011 = 2011.$

   A) 2013       B) 2011       C) 2       D) 1

   **1.** B

2. $(4+3) \times (5+2) \times (6+1) = (7) \times (7) \times (7) = 7^3.$

   A) $3 \times 7$   B) $7 \times 7$   C) $3^7$   D) $7^3$

   **2.** D

3. Ben finished wrapping 30 boxes at 1:30 PM. It took him $5 \times 30 = 150$ minutes = 2½ hours. He began at 11:00 AM.

   A) 10:30   B) 11:00   C) 11:30   D) 11:50

   **3.** B

4. $2\dfrac{3}{4} + 3\dfrac{4}{5} = \dfrac{11}{4} + \dfrac{19}{5} = \dfrac{55}{20} + \dfrac{76}{20} = \dfrac{131}{20} = 6\dfrac{11}{20}.$

   A) $5\dfrac{3}{5}$   B) $5\dfrac{11}{20}$   C) $6\dfrac{3}{5}$   D) $6\dfrac{11}{20}$

   **4.** D

5. The ones digit of the cube of 432 is the ones digit of $2^3$, which is 8.

   A) 8       B) 6       C) 4       D) 2

   **5.** A

6. Each choice has been rounded to the nearest whole number. The remainder when divided by 3 is shown.

   A) 14, R=2   B) 16, R=1   C) 16, R=1   D) 18, R=0

   **6.** A

7. Since $351 \div 3 = 117$, and $117 = 3 \times 3 \times 13$, another prime factor is 13.

   A) 7       B) 13       C) 39       D) 117

   **7.** B

8. Since $90° - 28° = 62°$, choice C is correct. Note: The smallest angle must be 45° or less.

   A) 1°   B) 30°   C) 62°   D) 91°

   **8.** C

9. The least common multiple of $45 = 3 \times 3 \times 5$ and $105 = 3 \times 5 \times 7$ is $315 = 3 \times 3 \times 5 \times 7$. That's 5 hours and 15 minutes after 1 PM, which is 6:15 PM.

   A) 3:30 PM   B) 4:30 PM   C) 6:15 PM   D) 7:15 PM

   **9.** C

10. If 6 cronks = 14 crunks, then 3 cronks = 7 crunks, and 9 cronks = 21 crunks.

    A) 24       B) 21       C) 20       D) 17

    **10.** B

11. $20 \div 0.4 = 50.$

    A) $\dfrac{1}{2}$       B) 8       C) 25       D) 50

    **11.** D

12. (Area of square) $-$ (area of triangle) $= 36 - 8 = 28.$

    A) 8   B) 16   C) 28   D) 36

    **12.** C

13. The only two prime numbers between 80 and 90 are 83 and 89.

    A) 1   B) 2   C) 3   D) 4

    **13.** B

Go on to the next page ))))➡ 7

14. There are 20 chess pieces. There are
20 − 12 = 8 black pieces. The ratio of
black pieces to all pieces is 8:20 = 2:5.

   A) 1:2    B) 1:4    C) 2:3    D) 2:5

15. The cube of 2 is 8, the square of 2 is 4,
and 8 + 4 = 12.

   A) 2      B) 4      C) 16     D) 64

16. My 7 Larry Rotter books of 300 pages each have 2100 total pages.
My 5 Chronicles of Blarnia books of 324 pages each have 1620 total
pages. The average page count is (2100 + 1620)÷12 = 310.

   A) 310        B) 312        C) 314        D) 316

17. The product of the first 3 positive perfect squares
is 1 × 4 × 9 = 36. The reciprocal is choice B.

   A) $\frac{1}{576}$    B) $\frac{1}{36}$    C) 36    D) 576

18. Each piece is 12 m ÷ 4 = 3 m long. The circumference
of each wheel must also be 3 m. Since C = πd, we
have 3 m = πd. Therefore, d = (3/π) m.

   A) 6π    B) 3π    C) $\frac{3}{\pi}$    D) $\frac{3}{2\pi}$

19. The sum of four consecutive whole numbers is 110.  Since 110÷4 =
27.5, the numbers are 26, 27, 28, and 29. The desired sum is 26 + 29.

   A) 53        B) 55        C) 57        D) 58

20. If the measure of one angle of a triangle is greater than 90°, the other
angles must each have measures less than 90°. They are both acute.

   A) acute      B) obtuse      C) right      D) scalene

21. The area of the entire figure is 64 and its height
is 4. Thus, the base of this figure is 64 ÷ 4 = 16.
The perimeter is 2 × (4+16) = 40.

   A) 52        B) 48        C) 44        D) 40

22. 1 440 000 mins. ÷ 60 = 24 000 hrs. = 1000 days;  so it's 10 AM again.

   A) 10 AM      B) 11 AM      C) 10 PM      D) 11 PM

23. 6:8.4 = (6 × 10):(8.4 × 10) = 60:84 = (60÷12):(84÷12) = 5:7.

   A) 2:4.4        B) 3:5        C) 5:7        D) 6.8:4

24. The sum of the measures of two unequal angles of a parallelogram
is 180°. So these two angles must have measures of 30° and 150°.

   A) 100°        B) 120°        C) 150°        D) 160°

*Go on to the next page* ))))➡ **7**

| | |
|---|---|
| 25. Joy walks 60 m in 180 sec., so she walks 1 m in 3 secs. She slides 9 times faster, so she slides 9 m in 3 secs. Thus, she slides 90 m in 30 seconds.<br><br>A) 90 m   B) 180 m   C) 270 m   D) 810 m | 25.<br><br>A |
| 26. If $\frac{2}{3} \times \#$ is $\frac{1}{2}$, then $\frac{1}{3} \times \#$ is $\frac{1}{4}$. Thus, $\frac{1}{6} \times \#$ is $\frac{1}{8}$.<br><br>A) $\frac{1}{8}$    B) $\frac{2}{9}$    C) $\frac{3}{16}$    D) $\frac{4}{15}$ | 26.<br><br>A |
| 27. 4 hrs. = 240 mins. = 14 400 secs.;  4 hrs. is $(14\,400 \div 24) \times 100\% = 60\,000\%$.<br><br>A) 600%       B) 3600%       C) 60 000%       D) 360 000% | 27.<br><br>C |
| 28. Let Cal's height be 100 units. Then Bo's height is 75 units and Abe's height is 105 units. Abe's height is 105 percent of Cal's height.<br><br>A) 25%       B) 85%       C) 95%       D) 105% | 28.<br><br>D |
| 29. $18^{180} = 2^{180} \times (3^2)^{180} = 2^{180} \times 3^{360}$, and $12^{360}$ is divisible by $3^{360}$.<br><br>A) $12^{120}$       B) $12^{180}$       C) $12^{240}$       D) $12^{360}$ | 29.<br><br>D |
| 30. Each of the 51 even integers between 19 and 121 is 1 less than each of the 51 odd integers between 20 and 122. Their sums differ by 51.<br><br>A) 1          B) 51          C) 100          D) 101 | 30.<br><br>B |
| 31. Since 30 = 5×6, 70 = 5×14, and 84 = 6×14, the box's dimensions are 5, 6, and 14. The volume is 5×6×14.<br>A) 184     B) 368     C) 420     D) 176 400 | 31.<br><br>C |
| 32. The consecutive even integers are shown for A, B, D.<br><br>A) -2, 0, 2, 4   B) 0, 2, 4, 6   C) 16   D) 2, 4, 6, 8 | 32.<br><br>C |
| 33. If the product is divided by 210, the remainder is 0 since 210 = 2×3×5×7 = product of the first 4 primes.<br><br>A) 0       B) 3       C) 7       D) 21 | 33.<br><br>A |
| 34. We'll use ROY G BIV. With R or G, Amy can pick 6 pairs: OY, OI, OV, YI, YV, or IV. That's 12 so far. Without R or G, Amy can pick OYI, OYV, OIV, and YIV. In all, Amy can pick 16 color combinations.<br><br>A) 10        B) 16        C) 20        D) 24 | 34.<br><br>B |
| 35. The sum of the lengths of the two other sides is greater than 9. The perimeter is greater than 9+9 = 18. Only choice D is greater than 18.<br><br>A) 11        B) 16        C) 18        D) 38 | 35.<br><br>D |

*The end of the contest*  🖎 **7**

**Visit our Web site at http://www.mathleague.com**

92

# 8th Grade Solutions

## 2006-2007 through 2010-2011

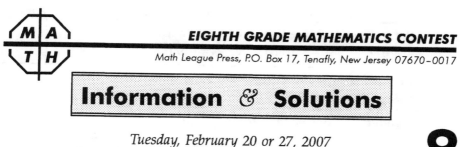

# Information & Solutions

*Tuesday, February 20 or 27, 2007*

## Contest Information

- **Solutions** Turn the page for detailed contest solutions (written in the question boxes) and letter answers (written in the *Answers* column to the right of each question).

- **Scores** Please remember that *this is a contest, not a test*—and there is no "passing" or "failing" score. Few students score as high as 30 points (75% correct). Students with half that, 15 points, *deserve commendation!*

- **Answers & Rating Scale** Turn to page 143 for the letter answers to each question and the rating scale for this contest.

| | |
|---|---|
| 1. Of the following, choice D has the least value.<br>A) $\frac{2}{3} = 0.666...$  B) $\frac{3}{5} = 0.600$   C) $\frac{4}{7} = 0.571...$  D) $\frac{5}{9} = 0.555...$ | 1.<br>D |
| 2. 27 fifty-cent stamps cost \$13.50. I got \$6.50 change from \$20.<br>A) \$18.65      B) \$13.50      C) \$10.35      D) \$6.50 | 2.<br>D |
| 3. The product is 0 since one factor is $\frac{0}{1} = 0$.<br>A) 1       B) $\frac{1}{5}$       C) 0       D) $-\frac{1}{5}$ | 3.<br>C |
| 4. July 1 is Friday, so the 8th, 15th, 22nd, and 29th are Fridays too. July 31 is 2 days later, a Sunday.<br>A) Saturday   B) Sunday<br>C) Monday    D) Tuesday | 4.<br>B |
| 5. $(10 \times 0.1) \times (100 \times 0.01) \times (1000 \times 0.001) = 1 \times 1 \times 1 = 1.$<br>A) 1      B) 0.1      C) 0.01      D) 0.001 | 5.<br>A |
| 6. The 99 smallest positive integers are 1, 2, 3, . . . , 99. The 99 largest negative integers are –1, –2, –3, . . . , –99. The sum is 0.<br>A) 0           B) 4950      C) 9900      D) 10 000 | 6.<br>A |
| 7. Since I saved $90 = 3 \times 30$ nickels, I saved $2 \times 30 = 60$ dimes.<br>A) 15         B) 30         C) 60         D) 90 | 7.<br>C |
| 8. $\frac{1}{2} + \frac{3}{4} = \frac{2}{4} + \frac{3}{4} = \frac{5}{4}.$<br>A) 4         B) 6         C) 7         D) 12 | 8.<br>A |
| 9. $m\angle O > 90°$ and $m\angle A < 90°$, so $m\angle O - m\angle A$ can *never* equal 0°.<br>A) 90°        B) 89°        C) 1°        D) 0° | 9.<br>D |
| 10. 120 times in 15 minutes averages $120 \div 15 = 8$ times each minute, or $3 \times 8 = 24$ times in 3 minutes.<br>A) 8      B) 12      C) 24      D) 30 | 10.<br>C |
| 11. $(1/10) - (1/100) = 0.1 - 0.01 = 0.10 - 0.01 = 0.09.$<br>A) 0.9      B) 0.09      C) 0.10      D) 10 | 11.<br>B |
| 12. Given $= 100\,000^2 \div (100 \times 1000) = 100\,000.$<br>A) 100 000     B) 10 000     C) 10     D) 1 | 12.<br>A |
| 13. Odd factors of 2007: greatest = 2007; least = 1; difference = 2006.<br>A) 666        B) 668        C) 2004        D) 2006 | 13.<br>D |
| 14. 9 tenths + 9 hundredths – 9 thousandths = $0.9 + 0.09 - 0.009 = 0.981$.<br>A) 0.9901     B) 0.981     C) 0.99     D) –0.901 | 14.<br>B |
| 15. $1 + 1/1 = 2$, so not B or C; and $-3 + 1/(-3) = -10/3$, so not D.<br>A) 0        B) 2        C) positive    D) negative | 15.<br>A |

*Go on to the next page* ▸ **8**

| | |
|---|---|
| 16. Divisibility by 12 and 21 implies divisibility by 4, 7, and $4\times7 = 28$.<br>A) 28 B) 33 C) 36 D) 63 | 16.<br>A |
| 17. The average of *any* two numbers always equals half their sum.<br>A) sum B) product C) quotient D) difference | 17.<br>A |
| 18. Factoring, $2^2 + 2^2 \times 2^2 + 2^2 \times 2^2 = 2^2 \times (1 + 2^2 + 2^2) = 2^2 \times 9$.<br>A) 5 B) 6 C) 9 D) 16 | 18.<br>C |
| 19. Since $120 \times 30 = 3600 = 60^2$, choice D is correct. For no other choice is the product a square.<br>A) 6 B) 10 C) 15 D) 30 | 19.<br>D |
| 20. $0.1 + 0.01 + 0.001 = 0.111 =$ 111 thousandths.<br>A) 1 B) 11 C) 100 D) 111 | 20.<br>D |
| 21. Half of $0.5\% = 0.25\% = 5 \times 0.05\%$.<br>A) 5% B) 0.05% C) 0.005% D) 0.0005% | 21.<br>B |
| 22. In 15 hours, at 1 cm/hr., it burns out; at 0.5 cm/hr., it burns halfway.<br>A) 3.75 hours B) 7.5 hours C) 15 hours D) 22.5 hours | 22.<br>C |
| 23. Top face + front face + side face = $(1 \times 2) + (1 \times 3) + (2 \times 3) = 11$. The sum of the areas of all six faces is 22.<br>A) 6 B) 11 C) 22 D) 36 | 23.<br>C |
| 24. $\dfrac{1}{1 + \frac{1}{2}} = \dfrac{1}{\frac{3}{2}} = 1 \div \frac{3}{2} = 1 \times \frac{2}{3} = \frac{2}{3}$.<br>A) 1 B) 2 C) 3 D) 4 | 24.<br>C |
| 25. $(1/8) \times (1/8) = 1/64 = (1/4) \times (1/16) =$ one-fourth of one-sixteenth.<br>A) one-sixteenth B) one-fourth C) one-third D) one-half | 25.<br>B |
| 26. Each side of a regular pentagon is 1/5 (or 20%) of its perimeter.<br>A) pentagon B) hexagon C) octagon D) decagon | 26.<br>A |
| 27. $30 = 2 \times 3 \times 5$ has these factors: 1, 2, 3, 5, 6, 10, 15, 30.<br>A) 4 B) 6 C) 7 D) 8 | 27.<br>D |
| 28. If the average of 9 numbers is 10, their sum is $9 \times 10 = 90$.<br>A) 14 B) 19 C) 80 D) 90 | 28.<br>D |
| 29. Since 1 CD at full price costs $16, 4 CDs on sale cost $3 \times \$16 = \$48$. Thus, 1 CD on sale costs $12, and 9 CDs on sale cost $9 \times \$12 = \$108$.<br>A) $144 B) $108 C) $72 D) $48 | 29.<br>B |

*Go on to the next page* ⫸ **8**

97

| | | |
|---|---|---|
| 30. | Each side of the square is 12. Subtract $36\pi$ (circle's area) from 144 (square's area) to get choice D.<br>A) $36-12\pi$ B) $144-12\pi$ C) $36-36\pi$ D) $144-36\pi$ | 30.<br>D |
| 31. | The average value of all 1999 quotients is the value of the middle quotient, whose value is $1000 \div 2 = 500$.<br>A) 498.5 B) 499 C) 499.5 D) 500 | 31.<br>D |
| 32. | $a = 3, b = 4$, so $3\triangle4 = 4^3+(3\times4) = 64+12 = 76$.<br>A) 76 B) 84 C) 88 D) 93 | 32.<br>A |
| 33. | Since your piggy bank had three times as many coins as mine, we count your coins three times and my coins once. Taken together, the % that are dimes is $(70\%+75\%+75\%+75\%)\div4 = 73.75\%$.<br>A) 72.5 B) 73.25 C) 73.75 D) 74 | 33.<br>C |
| 34. | The lcm of 1, 2, 3, 4, 5, 6, 7, 8, 9, and 10 is $5\times7\times8\times9 = 2520$.<br>A) 3 628 800 B) 7560 C) 2520 D) 1260 | 34.<br>C |
| 35. | $1000^{1001}-1000^{1000} = 1000^{1000}\times(1000^1-1) = 1000^{1000}\times999$.<br>A) 1000 B) $999^{1000}$ C) $1000\times1001$ D) $999\times1000^{1000}$ | 35.<br>D |
| 36. | The sum of 5 primes is odd, so all 5 primes must be odd. Since 1 is not a prime, the sum of the 5 primes could be $3+5+7+11+13 = 39$.<br>A) 39 B) 35 C) 27 D) 25 | 36.<br>A |
| 37. | If the measures of the angles of a triangle are 3 different numbers, each side has a different length.<br>A) scalene B) obtuse C) isosceles D) equilateral | 37.<br>A |
| 38. | If the square root of the perimeter of the triangle is 6, then the perimeter is $6^2 = 36$, and the length of each side is $36\div3 = 12$.<br>A) 8 B) 9 C) 12 D) 16 | 38.<br>C |
| 39. | The sum of the digits of $10^{200}+2 = 100\ldots002$ is $1+2 = 3$.<br>A) $10^{200} + 1$ B) $10^{200} + 2$ C) $10^{200} + 3$ D) $10^{200} + 4$ | 39.<br>B |
| 40. | Coloring the squares red, yellow, and blue, as illustrated, shows how to use 3 colors.<br>A) 2 B) 3 C) 4 D) 5 | 40.<br>B |

*The end of the contest* **8**

**Visit our Web site at http://www.mathleague.com**

98

### EIGHTH GRADE MATHEMATICS CONTEST

Math League Press, P.O. Box 17, Tenafly, New Jersey 07670–0017

# Information & Solutions

*Tuesday, February 19 or 26, 2008*

## Contest Information

**8**

- **Solutions** Turn the page for detailed contest solutions (written in the question boxes) and letter answers (written in the *Answers* column to the right of each question).

- **Scores** Please remember that *this is a contest, not a test*—and there is no "passing" or "failing" score. Few students score as high as 30 points (75% correct). Students with half that, 15 points, *deserve commendation!*

- **Answers & Rating Scale** Turn to page 144 for the letter answers to each question and the rating scale for this contest.

1. $(10 + 70) + (20 + 60) + (30 + 50) = 80 + 80 + 80 = 80 \times 3.$
   A) 3    B) 4    C) 6    D) 10

   1. A

2. Check only divisibility by 16. This assures divisibility by 2, 4, and 8.
   A) 1624    B) 2461    C) 3218    D) 4816

   2. D

3. $0.05 \times 0.01$ has a total of 4 decimal places, as does $0.5 \times 0.001$.
   A) 01    B) 0.01    C) 0.001    D) 0.0001

   3. C

4. Since the perimeter is 1, the average of the side-lengths is $\frac{1}{3}$.
   A) 1    B) $\frac{1}{6}$    C) $\frac{1}{3}$    D) 3

   4. C

5. If Fido scored 6 points every 12 minutes, then he scored $6 \times 6$ points $= 36$ points in $6 \times 12$ minutes $= 72$ minutes $= 1.2$ hours.
   A) 1    B) 1.2    C) 1.5    D) 2

   5. B

6. Since 100 months = 8 years + 4 months, Fido, who was more than 8 years younger, was 5.
   A) 4    B) 5    C) 6    D) 7

   6. B

7. $\frac{1}{2} \times \frac{1}{2} + \frac{1}{2} \times \frac{1}{2} = \frac{1}{4} + \frac{1}{4} = \frac{1}{2} \times 1.$    A) $\frac{1}{8}$  B) $\frac{1}{4}$  C) $\frac{1}{2}$  D) 1

   7. D

8. 100 thousandths $-$ 5 hundredths $= 0.100 - 0.050 = 0.050 = 5/100$.
   A) $\frac{5}{100}$    B) $\frac{20}{100}$    C) $\frac{10}{100}$    D) $\frac{95}{1000}$

   8. A

9. Write each with 5 decimal places. Only $0.20800 > 0.20080$.
   A) 0.20800    B) 0.20000    C) 0.02080    D) 0.20008

   9. A

10. The sum of all 3 angles is 180°, so at most one is 90° or more.
    A) 0    B) 1    C) 2    D) 3

    10. C

11. $\frac{22+2}{2} + \frac{33+3}{3} + \frac{44+4}{4} = (11+1) \times 3 = 33+3.$
    A) 30+3  B) 33+3  C) 50+5  D) 55+5

    11. B

12. Yogi hibernates $\frac{1}{3}$ of $\frac{3}{5}$ of every year. That is $\frac{1}{3} \times \frac{3}{5} = \frac{1}{5} = 20\%$ of every year.
    A) 80  B) 50  C) 40  D) 20

    12. D

13. $(3 \times 1)^2 + (3 \times 2)^2 + (3 \times 3)^2 = 3^2 \times (1^2 + 2^2 + 3^2)$.
    A) 5    B) 6    C) 14    D) 15

    13. C

14. I have $4 \times 85 = 340$. To get $5 \times 88 = 440$, I'll need $440 - 340 = 100$.
    A) 91    B) 96    C) 98    D) 100

    14. D

15. Since $333\,333\,333 = 3 \times (111\,111\,111)$, it's divisible by 111.
    A) 11    B) 33    C) 111    D) 3333

    15. C

*Go on to the next page* ▷ **8**

| | |
|---|---|
| 16. $\frac{2!\times 3!\times 4!}{2\times 3\times 4}=\frac{2!\times 3!\times 4\times 3\times 2\times 1}{4\times 3\times 2\times 1}=2!\times 3!.$<br>A) 1 B) 4! C) $2!\times 3!$ D) $2\times 3\times 4$ | 16.<br>C |
| 17. Since $5n$ is always divisible by 5, $5n+1$ is never divisible by 5.<br>A) $n+5$ B) $3n+4$ C) $4n+3$ D) $5n+1$ | 17.<br>D |
| 18. To 2 (a prime), add 999 odd primes. Even + odd = odd.<br>A) 0 B) 1 C) 2 D) 3 | 18.<br>B |
| 19. The choice closest to its reciprocal is the choice closest to 1.<br>A) 0.01 B) 0.1 C) 1.01 D) 1.1 | 19.<br>C |
| 20. Since 1234 ÷ 24 = 51 with remainder 10, 1234 hours after midnight is the same as 10 hours after midnight. That's 10 A.M.<br>A) 10 A.M. B) noon C) 10 P.M. D) midnight | 20.<br>A |
| 21. A square with perimeter 24 has area $6^2=36$.<br>A) 24 B) 25 C) 36 D) 144 | 21.<br>C |
| 22. The units' digit of every power of 3 is odd.<br>A) 1 B) 3 C) 6 D) 9 | 22.<br>C |
| 23. $(999\,999\times 999\,999)-(999\,999\times 1)=999\,999\times(999\,999-1)$.<br>A) $1\,000\,000\times 999\,998$ B) $999\,999\times 999\,998$<br>C) $999\,999\times 1$ D) $999\,998\times 999\,998$ | 23.<br>B |
| 24. The perimeter of a rectangle $=2(\ell+w)$. When $\ell$ and $w$ are integers, $2(\ell+w)$ is even.<br>A) even B) odd C) prime D) > 4 | 24.<br>A |
| 25. Only 30 kids sing and/or play the drums. Since $24+16=40$, 10 kids must do both. The ratio of the number who do both to the number who do neither is 10:10 = 1:1.<br>A) 1:4 B) 3:5 C) 4:5 D) 1:1 | 25.<br>D |
| 26. 140 hooves ÷ (4 hooves per animal) = 35 animals. Since 2/5 of the animals are bulls, the number of bulls is $(2/5)\times 35=14$.<br>A) 35 B) 28 C) 21 D) 14 | 26.<br>D |
| 27. $\frac{1}{2+\frac{1}{2}}=\frac{1}{\frac{5}{2}}$, whose reciprocal is $\frac{5}{2}$. A) $\frac{2}{2+1}$ B) $\frac{2+1}{2}$ C) $\frac{2}{5}$ D) $\frac{5}{2}$ | 27.<br>D |
| 28. $(2\ell\times 0.02\text{ fat})+(3\ell\times 0.03\text{ fat})=(0.04+0.09)\ell$ fat per $5\ell=2.6\%$ fat.<br>A) 2.5% B) 2.6% C) 5% D) 6% | 28.<br>B |
| 29. If I start with 3, my 50th number is 150. If I start with 2, it's 149.<br>A) 149 B) 150 C) 151 D) 152 | 29.<br>A |

| | |
|---|---|
| 30. As shown below, the sum can be 14, 30, or 102, but not 52.<br>A) $14 = 5+5+4$  B) $30 = 25+4+1$   C) 52      D) $102 = 100+1+1$ | 30.<br>C |
| 31. Since $5 = 2+3$, $7 = 2+5$, and $8 = 3+5$, the numbers 5, 7, and 8 can be written as a sum of two primes. [Note: 1 is not prime.]<br>A) 3   B) 2   C) 1   D) none | 31.<br>A |
| 32. Right now, Pat is 8 years younger than Lee. In 5 years, Pat will be 8 and Lee will be 16. Lee is $16-5 = 11$ right now.<br>A) 16  B) 11  C) 5  D) 3 | 32.<br>B |
| 33. $0.02\times0.02 = 0.0004 = 0.04\times0.01$.<br>A) 10          B) 1          C) 0.1          D) 0.01 | 33.<br>D |
| 34. If the surface area is 24, each face's area is 4 and the length of each edge is 2. The volume is $2^3 = 8$. Each small cube's volume is 1.<br>A) 1          B) 3          C) 6          D) 8 | 34.<br>A |
| 35. If $a$ and $c$ are $< 0$, $a^3$ and $c^5$ are $< 0$, but their product is $> 0$.<br>A) $a$          B) $c$          C) both $a$ & $c$  D) both $b$ & $c$ | 35.<br>C |
| 36. $(2\times2)^8 = 2^8\times2^8 = 2^{16}$. $(2\times2\times2)^4 = 2^4\times2^4\times2^4 = 2^{12}$. GCF is $2^{12}$.<br>A) $2^{64}$          B) $2^{16}$          C) $2^{12}$          D) $2^4$ | 36.<br>C |
| 37. After Dad bought 3/5 of the fish that I caught, he gave away 1/4 of these fish and kept 3/4 of them. The fraction of my fish that Dad *kept* was 3/4 of 3/5 = 9/20.<br>A) $\frac{3}{20}$ B) $\frac{4}{20}$ C) $\frac{7}{20}$ D) $\frac{9}{20}$ | 37.<br>D |
| 38. The only 2-digit multiple of 5 that's 5 times the sum of its digits is 45.<br>A) 10  B) 20  C) 30  D) 40 | 38.<br>B |
| 39. The 8 factors are $1\times9, 2\times9, 3\times9, 5\times9, 6\times9, 10\times9, 15\times9,$ and $30\times9$.<br>A) 7          B) 8          C) 29          D) 30 | 39.<br>B |
| 40. Label the squares as shown. There are only 6 possible paths. These paths are 1-2-3-6-9, 1-2-5-6-9, 1-2-5-8-9, 1-4-5-6-9, 1-4-5-8-9, and 1-4-7-8-9.<br>A) 4          B) 6          C) 8          D) 12 | 40.<br>B |

*The end of the contest* ✍ **8**

**Visit our Web site at http://www.mathleague.com**

# Information & Solutions

*Tuesday, February 17 or 24, 2009*

## Contest Information

**8**

- **Solutions** Turn the page for detailed contest solutions (written in the question boxes) and letter answers (written in the *Answers* column to the right of each question).

- **Scores** Please remember that *this is a contest, not a test*—and there is no "passing" or "failing" score. Few students score as high as 30 points (75% correct). Students with half that, 15 points, *deserve commendation!*

- **Answers & Rating Scale** Turn to page 145 for the letter answers to each question and the rating scale for this contest.

1. $2\sqrt{25} - 2\sqrt{16} = 2\times5 - 2\times4 = 10 - 8 = 2$.
A) 2 B) 3 C) 4 D) 6
1. A

2. The closest we can get is $0.550 - 0.549 = 0.001$.
A) 0.49 B) 0.509 C) 0.549 D) 0.6
2. C

3. $1+\frac{1}{3}+2+\frac{2}{3}+3+\frac{3}{3} = 1+2+3+\left(\frac{1}{3}+\frac{2}{3}+\frac{3}{3}\right) = 6+\frac{6}{3} = 6+2 = 8$.
A) 9 B) 8 C) 7 D) 6
3. B

4. The greatest factor of $39\times49\times59$ is $39\times49\times59$, an odd number.
A) 9 B) prime C) even D) odd
4. D

5. 99 hundredths − 99 thousandths = 0.990 − 0.099 = 0.891.
A) 0.891 B) 0.81 C) 0.01 D) 0.001
5. A

6. Each choice has been replaced by its reciprocal.
A) $-\frac{4}{3}$ B) $\frac{6}{5}$ C) $-\frac{8}{7}$ D) $\frac{10}{9}$
6. B

7. Since $8\times12 = 96$, there are 96 containers. The only multiple of 96 among the choices is $3\times96 = 288$.
A) 280 B) 284 C) 288 D) 292

7. C

8. $11\times365 = 4015$, so 4000 days is about half a month before June 1.
A) April B) May C) June D) July
8. B

9. Any percent of 0% is 0%.
A) 0 B) 100 C) 100% D) 240 000%
9. A

10. The length of each side is $\frac{3}{4}\div4 = \frac{3}{16}$, and the area is $\left(\frac{3}{16}\right)^2$.
A) $\frac{3}{4}$ B) $\frac{3}{16}$ C) $\frac{9}{4}$ D) $\frac{9}{256}$
10. D

11. If 120% of my hourly wages is \$120, 100% of my hourly wages is \$100. That represents 8 hours of work for \$12.50 per hour.
A) \$10 B) \$12 C) \$12.50 D) \$15
11. C

12. 250% = 2.50 = 2.5 = 5/2.
A) $\frac{1}{4}$ B) $\frac{2}{5}$ C) $\frac{5}{2}$ D) 25
12. C

13. The l.c.m. of 11, $2\times11$, $3\times11$, and $4\times11$ is $3\times4\times11 = 132$.
A) 66 B) 88 C) 99 D) 132
13. D

14. The sum of the 102 integers is 51. Their average is $51\div102 = 0.5$.
A) 0 B) 0.5 C) 1 D) 50
14. B

Go on to the next page ➡ 8

| | | |
|---|---|---|
| 15. | Beginning Sunday, in 6 days I read $2 + 4 + 8 + 16 + 32 + 64 = 126$ pages. I read my 100th page on Friday.<br>A) Friday    B) Saturday    C) Monday    D) Tuesday | 15.<br>A |
| 16. | Al took 25% of my books and Ed took *half* the remaining 3/4, so Ed took 3/8 of my books. Ed's *half* equals my half, 30, so 3/8 of my books is 30, 8/8 is 80, and 25% of 80 is 20.<br>A) 15    B) 20    C) 60    D) 80 | 16.<br>B |
| 17. | On a number line, –0.1 is furthest right.<br>A) –0.1    B) $(-10)^3$    C) –100    D) $-\sqrt{100}$ | 17.<br>A |
| 18. | $3 \times 6 \times 8 = 3 \times 6 \times 2 \times 4 = (2 \times 3 \times 6) \times 4$.<br>A) 1.5    B) 2    C) 3    D) 4 | 18.<br>D |
| 19. | 10:45 A.M. is 150 mins. past 8:15 A.M. & 300 mins. before 3:45 P.M.<br>A) 4 A.M.    B) 10:45 A.M.    C) 12 P.M.    D) 1:15 P.M. | 19.<br>B |
| 20. | Since $600 = 24 \times 25$, the sum of the integers is $24 + 25 = 49$.<br>A) 48    B) 49    C) 50    D) 60 | 20.<br>B |
| 21. | Since $\pi r^2 = 8 \times (2\pi r) = 16\pi r$, $r^2 = 16r$, and $r = 16$.<br>A) 2    B) 4    C) 8    D) 16 | 21.<br>D |
| 22. | Sum $= \frac{1}{1} + \frac{1}{2} + \frac{1}{2} + \frac{1}{3} + \frac{1}{3} + \frac{1}{3} + \frac{1}{4} + \frac{1}{4} + \frac{1}{4} + \frac{1}{4} = 1+1+1+1 = 4$.<br>A) 4    B) 3    C) 2    D) 1 | 22.<br>A |
| 23. | 20 bowls = 8 cups, and 15 mugs = 20 bowls; 8 cups = 15 mugs.<br>A) 20 mugs    B) 16 mugs    C) 15 mugs    D) 12 mugs | 23.<br>C |
| 24. | $24 = 3 \times 8$ and $124 = 31 \times 4$. The common factors are 1, 2, and 4.<br>A) 1    B) 2    C) 3    D) 4 | 24.<br>C |
| 25. | In dollars, the average cost of these 30 flights was $(10 \times 95 + 20 \times 86)/30 = 2670/30 = 89$.<br>A) $89    B) $90    C) $91    D) $92 | 25.<br>A |
| 26. | $(-1)^2 - (-1^2) = 1 - (-1) = 1 + 1 = 2$.<br>A) –2    B) –1    C) 1    D) 2 | 26.<br>D |
| 27. | $\sqrt{36} = 6$ is half of $12 = \sqrt{144}$.<br>A) $\sqrt{18}$    B) $\sqrt{72}$    C) $\sqrt{128}$    D) $\sqrt{144}$ | 27.<br>D |
| 28. | By trial, only $18 = 2 \times (1+8)$ is twice the sum of its digits.<br>A) 4    B) 2    C) 1    D) 0 | 28.<br>C |

*Go on to the next page* ⟹ **8**

29. Using the Pythagorean Theorem, $13^2-5^2 = 144 = 12^2$. The area is 5 cm $\times$ 12 cm = 60 cm$^2$.

   A) 34    B) 36    C) 60    D) 65

   29. C

30. If the area is $16\pi^2$, the side-length is $4\pi$ and the perimeter is $16\pi$.

   A) $\frac{\pi^2}{16}$    B) $\frac{16}{\pi^2}$    C) 4    D) $16\pi^2$

   30. D

31. Since 24 is divisible by $2^3 = 8$, $24^{30}$ is divisible by $(2^3)^{30} = 2^{90}$.

   A) $18^{36}$  B) $24^{30}$  C) $30^{24}$  D) $36^{18}$

   31. B

32. If the shortest side were 7, the sum of the other 2 sides would need to be at least 14, not the actual 11.

   A) 4  B) 5  C) 6  D) 7

   32. D

33. The circle can cross each side of the rectangle twice, for a total of 8.

   A) 8    B) 7    C) 6    D) 4

   33. A

34. For the average to be 50°, the sum must be 100°, and 10°+90° = 100°.

   A) 30°    B) 40°    C) 50°    D) 90°

   34. C

35. If $\left(\frac{a}{b}\right)^{-c}$ equals $\left(\frac{b}{a}\right)^c$, then $\left(\frac{2}{3}\right)^{-4} = \left(\frac{3}{2}\right)^4 = \frac{81}{16}$.

   A) $\frac{81}{16}$    B) $\frac{16}{3}$    C) $\frac{3}{16}$    D) $\frac{2}{81}$

   35. A

36. $(\pi r^2)/(2\pi r)^2 = (\pi r^2)/(4\pi^2 r^2) = 1/(4\pi)$.

   A) $\frac{1}{4\pi}$    B) $\frac{1}{4\pi^2}$    C) $\frac{1}{2\pi}$    D) $\frac{1}{4}$

   36. A

37. Every integer from 2 to 2010 is divisible by a prime less than 2010. [NOTE: 1 is *not* a prime.]

   A) 2010    B) 2009
   C) 1004    D) 1005

   37. B

38. The sum of 1024 fours is $4\times 1024 = 4\times 4^5 = 4^6$.

   A) $4^4$    B) $4^5$    C) $4^6$    D) $4^7$

   38. C

39. Reduce $\frac{24}{42}$ to get $\frac{4}{7}$. Multiply $\frac{4}{7}$ by $\frac{5}{5}$ to get $\frac{20}{35}$.

   A) 35    B) 45    C) 55    D) 65

   39. A

40. $(151+152+153+\ldots+200) = (150+1 + 150+2 + 150+3 + \ldots + 150+50) = 50\times 150 + (1+2+3+\ldots+50) = 7500+1275 = 8775$.

   A) 6275    B) 6600    C) 8375    D) 8775

   40. D

*The end of the contest* ✍ **8**

**Visit our Web site at http://www.mathleague.com**

# Information & Solutions

## 2009-2010 Annual 8th Grade Contest

*Tuesday, February 16 or 23, 2010*

**8**

### Contest Information

- **Solutions** Turn the page for detailed contest solutions (written in the question boxes) and letter answers (written in the *Answer Column* to the right of each question).

- **Scores** Please remember that *this is a contest, and not a test*—there is no "passing" or "failing" score. Few students score as high as 30 points (75% correct); students with half that, 15 points, *deserve commendation!*

- **Answers and Rating Scales** Turn to page 146 for the letter answers to each question and the rating scale for this contest.

| | |
|---|---|
| 1. Factor: 2010 = 2 × 3 × 5 × 67; or use divisibility tests.<br><br>A) 2      B) 3      C) 5      D) 7 | 1.<br><br>D |
| 2. 28% is equal to 28/100 = 7/25.<br><br>A) 2.8    B) $\frac{7}{25}$   C) 2800   D) $\frac{0.28}{100}$ | 2.<br><br>B |
| 3. $\frac{4}{5} - \frac{3}{20} = \frac{16}{20} - \frac{3}{20} = \frac{13}{20}$.<br><br>A) $\frac{1}{15}$   B) $\frac{2}{15}$   C) $\frac{13}{20}$   D) $\frac{12}{65}$ | 3.<br><br>C |
| 4. June has 30 days. Al worked at the pool for 0.2 × 30 = 6 days.<br><br>A) 3      B) 6      C) 9      D) 12 | 4.<br><br>B |
| 5. Of choices, only 110° and 120° are obtuse angles; 120° + 60° = 180° already, 110° + 60° + 10° works.<br><br>A) 30°    B) 90°    C) 110°   D) 120° | 5.<br><br>C |
| 6. 0.8 = 8/10 = 4/5.<br><br>A) $\frac{4}{5}$      B) $\frac{5}{4}$      C) $\frac{1}{8}$      D) $\frac{8}{100}$ | 6.<br><br>A |
| 7. 6000 seconds = 100 minutes = 1 hour, 40 minutes. It is now 11:40 AM.<br><br>A) 11:00 AM    B) 11:40 AM    C) 4:00 PM    D) 10:00 PM | 7.<br><br>B |
| 8. 2 − 13 − (−7) = 2 − 13 + 7 = −11 + 7 = −4.<br><br>A) −18      B) −11      C) −8      D) −4 | 8.<br><br>D |
| 9. $\sqrt{144}$ = 12; 5 less than 12 is 7, and 5 less than 7 is 2 = $\sqrt{4}$.<br><br>A) $\sqrt{4}$      B) $\sqrt{25}$      C) $\sqrt{49}$      D) $\sqrt{134}$ | 9.<br><br>A |
| 10. Try numbers: 100 × 1000 = 100000. This product has 6 digits.<br><br>A) 12      B) 8      C) 6      D) 5 | 10.<br><br>C |
| 11. At 2, the min. hand is 10 min. marks, or 10×6°, away from the hr. hand.<br><br>A) 2:00    B) 3:30    C) 6:00    D) 9:45 | 11.<br><br>A |
| 12. The 4 primes are 29, 31, 37, and 41.<br><br>A) 3      B) 4      C) 5      D) 6 | 12.<br><br>B |
| 13. $\frac{6 \times 5 \times 5 \times 5 \times 4 \times 5 \times 3 \times 5}{6 \times 5 \times 4 \times 3} = 5 \times 5 \times 5 \times 5 = 5^4$.<br><br>A) $5^4$    B) $5^3$    C) $5^2$    D) $5^1$ | 13.<br><br>A |
| 14. 0.3 × 40 = 0.4 × 30 = 40% of 30.<br><br>A) 200    B) 120    C) 60    D) 30 | 14.<br><br>D |
| 15. The equilateral triangle has perimeter 3×8 = 24. The length of a side of the square is 24÷4 = 6, so its area is 36.<br><br>A) 16      B) 24      C) 36      D) 64 | 15.<br><br>C |

*Go on to the next page* ))⟫ **8**

16. $\frac{1}{2} \div \frac{3}{8} = \frac{1}{2} \times \frac{8}{3} = \frac{8}{6} = \frac{4}{3}$.

   A) $\frac{3}{16}$   B) $\frac{2}{5}$   C) $\frac{5}{8}$   D) $\frac{4}{3}$

16.

D

17. $3^4 + 3^4 + 3^4 = 3 \times 3^4 = 3^1 \times 3^4 = 3^5$.

   A) $3^5$   B) $9^4$   C) $3^{12}$   D) $9^{12}$

17.

A

18. Eli's goal was to lift 3.5 kg during his workout, but he was able to lift only 3 kg. Eli lifted 3/3.5 = 6/7 of his goal weight.

   A) $\frac{1}{7}$   B) $\frac{2}{3}$   C) $\frac{3}{4}$   D) $\frac{6}{7}$

18.

D

19. As shown below, choice B is the largest.

   A) 0.02   B) 0.05   C) 0.04   D) 0.03

19.

B

20. Choose any number as a side-length of A. If you choose 2, A's perimeter is 8, B's is 4, and B's side-length is 1. The areas are 4 and 1.

   A) $\frac{1}{2}$       B) 2       C) 4       D) 8

20.

C

21. Multiply through by 8: 13.25:1 = (13.25 × 8):(1 × 8) = 106:8.

   A) 53       B) 106       C) 122       D) 150

21.

B

22. The larger a positive number is, the smaller its reciprocal is.

   A) $\frac{2}{5}$       B) $\frac{3}{7}$       C) $\frac{4}{3}$       D) $\frac{9}{4}$

22.

D

23. Since 165÷15 = 11, we have 11 groups of 15 students. They'll need 11 × 2 = 22 teachers.

   A) 11       B) 13       C) 17       D) 22

23.

D

24. Since 4 oranges cost the same as 1 apple, replace 4 oranges with 1 apple to get 4 apples cost $4. Thus 2 apples cost $4÷2 = $2.

   A) $1.00   B) $1.50   C) $2.00   D) $2.50

24.

C

25. It's 1005 99s in a row, each div. by 11.

   A) 10   B) 11   C) 12   D) 15

25.

B

26. 250% = 2.5, and 2.5 × 12 = 30.

   A) 12   B) 45   C) 70   D) 75

26.

A

27. $2.01 \times 10^{2009} = 2.01 \times 10^3 \times 10^{2006} = 2010 \times 10^{2006}$.

   A) $10^{2006}$       B) $10^{2007}$       C) $10^{2012}$       D) $10^{2013}$

27.

A

28. The only prime number that is divisible by 2 is 2 itself.

   A) zero       B) one       C) three       D) ten

28.

B

29. The large circle's area is $16\pi$, so its radius is 4. A radius of a small circle is 2. The area of each small circle is $2^2 \times \pi = 4\pi$. The shaded region's area is $16\pi - (4\pi + 4\pi) = 8\pi$.

   A) $4\pi$   B) $5\pi$   C) $6\pi$   D) $8\pi$

29.

D

| | |
|---|---|
| 30. $14 = 1^2 + 2^2 + 3^2$, $21 = 1^2 + 2^2 + 4^2$, and $35 = 1^2 + 3^2 + 5^2$.<br><br>A) 14    B) 21    C) 28    D) 35 | 30.<br><br>C |
| 31. 99 months from Jan. 1 is 3 months more than 8 years (96 months). That's April 1.<br><br>A) March B) April C) May D) June | 31.<br><br>B |
| 32. $1 + 2 + 4 + 8 + 16 + 32 = 63$.<br><br>A) 5    B) 30    C) 32    D) 63 | 32.<br><br>D |
| 33. Jim has 12 socks: 4 red, 4 black, and 4 blue. Choosing in the dark, he wants at least one matching pair of socks that are *not* red. He could begin by choosing 4 red, 1 black, and 1 blue. The next one he chooses, the 7th one, must be either a black or blue sock.<br><br>A) 2        B) 3        C) 6        D) 7 | 33.<br><br>D |
| 34. When a number less than 2 is multiplied by a number less than 1, the product remains less than 2.<br><br>A) 0.25    B) 1    C) 1.75    D) 2.25 | 34.<br><br>D |
| 35. $2 \lozenge 3 = (2+3) \times (3-2) = 5 \times 1 = 5$; $1 \lozenge 5 = (1+5) \times (5-1) = 6 \times 4 = 24$.<br><br>A) 35    B) 24    C) 2    D) 0 | 35.<br><br>B |
| 36. The sum of any 6 odd integers is always even.<br><br>A) 108    B) 111    C) 333    D) 345 | 36.<br><br>A |
| 37. The first 8 flowers average 24 petals each. That's $8 \times 24 = 192$ petals total. The next 12 average 34 petals each, or $12 \times 34 = 408$ petals total. The average is $(192+408) \div 20 = 30$.<br><br>A) 28    B) 29    C) 30    D) 31 | 37.<br><br>C |
| 38. Since $28 = 2^2 \times 7$, its prime factors are 2 and 7, so $2^2$ and $7^2$ are factors of the square.<br><br>A) 784    B) 49    C) 20    D) 12 | 38.<br><br>B |
| 39. Harry sold \$7 on the 1st day; \$$(7 + 1 \times 3)$ on the 2nd day; \$$(7 + 2 \times 3)$ on the 3rd day; ... ; \$$(7 + 30 \times 3)$ on the 31st day; and \$$(7 + 31 \times 3)$ on the 32nd day. Since he began on July 1, he first sold \$100 worth on August 1.<br><br>A) July 30        B) July 31        C) August 1        D) August 2 | 39.<br><br>C |
| 40. Sixth powers are both squares and cubes. They always have at least 7 divisors. An example is $2^6 = 8^2 = 4^3$. Its 7 divisors are $1, 2, 2^2, 2^3, 2^4, 2^5, 2^6$.<br><br>A) 7        B) 6        C) 5        D) 4 | 40.<br><br>A |

*The end of the contest* 🖎 **8**

**Visit our Web site at http://www.mathleague.com**

110

## Information & Solutions

# 2010-2011 Annual 8th Grade Contest

*Tuesday, February 15 or 22, 2011*

**8**

## Contest Information

- **Solutions**  Turn the page for detailed contest solutions (written in the question boxes) and letter answers (written in the *Answer Column* to the right of each question).

- **Scores**  Please remember that *this is a contest, and not a test*—there is no "passing" or "failing" score. Few students score as high as 28 points (80% correct); students with half that, 14 points, *deserve commendation!*

- **Answers and Rating Scales**  Turn to page 147 for the letter answers to each question and the rating scale for this contest.

| | |
|---|---|
| 1. I have $222. If my friend loans me $789, I have $1011. I need $2011 − $1011 = $1000.<br><br>A) $0   B) $100   C) $1000   D) $1111 | 1.<br>C |
| 2. Al finds a gold nugget weighing 500 grams. Two would weigh 1000 grams = 1 kg.<br><br>A) 1 kg  B) 10 kg  C) 1000 kg  D) 10 000 kg | 2.<br>A |
| 3. $6+(8×0)+(10×2) − (12×0) = 6+0+20−0 = 26.$<br>A) 0    B) 8     C) 26     D) 38 | 3.<br>C |
| 4. There are 19 positive integers less than 20. Of these, 4 are multiples of 4. Therefore, the desired probability is 4/19.<br><br>A) $\frac{4}{19}$      B) $\frac{1}{5}$      C) $\frac{5}{19}$      D) $\frac{1}{4}$ | 4.<br>A |
| 5. The square of an integer minus the cube of that integer is never odd.<br><br>A) $(-1)^2−(-1)^3 = 2$ B) $2^2 − 2^3 = $ -4    C) $2^2 − 2^3 = $ -4    D) odd | 5.<br>D |
| 6. $127 = 1×127$; 127 is prime, but 1 is not.<br><br>A) $85 = 5 × 17$    B) $94 = 2 × 47$    C) $119 = 7 × 17$    D) 127 | 6.<br>D |
| 7. $(10 × 10) × (5 × 20) × (4 × 25) × (2 × 50) × 100 = 100^4 × 100.$<br><br>A) 4        B) 5        C) $100^4$       D) $100^5$ | 7.<br>C |
| 8. The number of dimes in $100 is 1000. The number of quarters in $200 is 800. The ratio of dimes to quarters is 1000:800 = 5:4.<br><br>A) 5:4      B) 4:5      C) 5:2      D) 2:5 | 8.<br>A |
| 9. The cost of dinner was $60 after the 20% tip was added. Thus $60 is 120% of the cost. Dinner without the tip was $60÷1.20 = $50. The cost for each of us without the tip was $50÷4 = $12.50.<br><br>A) $12      B) $12.50      C) $16      D) $16.67 | 9.<br>B |
| 10. To build a wall takes 12 hrs. for 18 workers or 12×18 = 216 hrs. for 1 worker. It would take 12 workers 216 hrs.÷12 = 18 hrs.<br>A) 6    B) 8    C) 14    D) 18 | 10.<br>D |
| 11. 10:59 AM+1 min. is midway between 10:59 PM today and 10:59 PM+2 mins. tomorrow.<br><br>A) 10 AM B) 11 AM C) 10 PM D) 11 PM | 11.<br>B |
| 12. $625 = 5 × 5 × 5 × 5$ has the most factors of 5.<br><br>A) 125      B) 500      C) 625      D) 750 | 12.<br>C |
| 13. $0.08 ÷ 0.004 = 80 ÷ 4 = 20 = 2000\%.$<br><br>A) 5%      B) 20%      C) 500%      D) 2000% | 13.<br>D |

14. Of every 9 balloons, 2 are fancy balloons. Since $2/9 \times 621 = 138$, Mr. B. Loon has 138 fancy balloons.

A) 138      B) 183      C) 207      D) 483

14. A

15. $\dfrac{\frac{1}{2}}{\frac{1}{3}+\frac{1}{4}} = \frac{1}{2} \div (\frac{1}{3}+\frac{1}{4}) = \frac{1}{2} \div \frac{7}{12} = \frac{1}{2} \times \frac{12}{7} = \frac{12}{14} = \frac{6}{7}$.

A) $\dfrac{4}{7}$      B) $\dfrac{6}{7}$      C) $\dfrac{7}{6}$      D) $\dfrac{7}{4}$

15. B

16. Since $2^5 \times 10^{52} = 32 \times 10^{52}$ and $10^{52}$ has 53 digits (a "1" followed by 52 "0s"), the product has 54 digits.

A) 54      B) 55      C) 56      D) 57

16. A

17. 1–9 has 9 digits; 10–29 has 40 additional digits. The 50th digit is a 3.

A) 0      B) 3      C) 4      D) 9

17. B

18. 32's only prime factor, 2, is a factor 5 times.

A) $30 = 2\times3\times5$    B) $32 = 2^5$    C) $34 = 2\times17$    D) $2^23^2$

18. B

19. Each angle in an equilateral $\triangle$ has a measure of 60°. The measures of the angles in an isosceles right $\triangle$ are 45°, 45°, and 90°; $60° - 45° = 15°$.

A) 15°      B) 45°      C) 60°      D) 75°

19. A

20. $4^8 = (2^2)^8 = 2^{16}$ is a multiple of $8^4 = (2^3)^4 = 2^{12}$.

A) $4^4$      B) $4^8$      C) $8^4$      D) $8^8$

20. B

21. The sum *cannot* be 3 since any such number is divisible by 3.

A) 2      B) 3      C) 4      D) 5

21. B

22. The number of seconds in an hour divided by the number of minutes in an hour is $(60\times60)\div(60) = 60$.

A) 5      B) 12      C) 60      D) 1440

22. C

23. If I'm 16, my age is doubled to 32 in 16 yrs. and tripled to 48 in 16 more.

A) 8      B) 12      C) 16      D) 32

23. C

24. If the avg. is 10, the sum is 10×10. If one is 92 and the others 1s, the sum is 101.

A) 10      B) 50      C) 90      D) 92

24. D

25. There are at least 3 riders on each toboggan. The sum of 21 odd primes (not necessarily all different) must be an odd number.

A) 110      B) 112      C) 121      D) 122

25. C

26. Each beaver in a colony of 20 beavers cuts 14 logs for a dam for a total of 280. Each beaver in a colony of 40 beavers cuts 20 logs for a total of 800. The average number of logs cut is $(280 + 800) \div (20 + 40) = 18$.

    A) 18    B) 17    C) 16    D) 15

    26.

    A

27. An example: If $r = 10$, then $A = 100\pi$. When $a = 25\pi$, $r = 5$, which is 50% less than $r = 10$.

    A) 25%   B) 50%   C) 60%   D) 75%

    27.

    B

28. The reciprocal of the reciprocal of $\sqrt{2}$ is $\sqrt{2}$, and $\sqrt{2} \times \sqrt{2} = 2$.

    A) $\sqrt{2}$      B) 2      C) $2\sqrt{2}$      D) 4

    28.

    B

29. $3m{:}8a = 15m{:}40a$ and $5a{:}9p = 40a{:}72p$. For every 15 melons there are 72 pears, and $15m{:}72p = 600{:}2880$.

    A) 600      B) 1320      C) 1440      D) 2880

    29.

    D

30. $r \blacklozenge s$ means $r^2 - 2s$, so $3 \blacklozenge (4 \blacklozenge 5) = 3 \blacklozenge (4^2 - 2 \times 5) = 3 \blacklozenge 6 = (3^2 - 2 \times 6) = -3$.

    A) -9      B) -3      C) 0      D) 3

    30.

    B

31. The original pairs $(d,p)$ for each choice would be (8,2), (12,3), (28,7), and (32,8). After spending 2 dimes and getting 2 pennies, they'd be (6,3), (10,5), (26,9), and (30,10). So only choice D works.

    A) 8      B) 12      C) 28      D) 32

    31.

    D

32. Since the area of the semicircle is $\frac{1}{2} \times 2^2 \times \pi = 2\pi$, the area of the triangle is 16. The base of the triangle is 4. Since area of a triangle is $bh/2$, we have $16 = 4h/2$, so $h$ is 8. The greatest possible distance between two points of the figure is the length of the line segment shown, $8 + 2 = 10$.

    A) 10    B) 12    C) 14    D) 16

    32.

    A

33. Pat's drawer is 100 cm wide, 200 cm long, and 50 cm deep. Its capacity is $(100 \times 200 \times 50)$ cubic cm.

    A) 0.01   B) 100    C) 10 000   D) 1 000 000

    33.

    D

34. There are 20 houses from the 12th to the 31st. The range of the house numbers is $2131 - 2011$ or 120. Adjacent house numbers differ by $120 \div 20 = 6$. Go back by 6s from the 11th to the 1st: $2011 - 6 \times 10 = 1951$.

    A) 1901   B) 1945   C) 1951   D) 1968

    34.

    C

35. The integers less than 2011 with an odd number of factors are the 44 perfect squares: 1, 4, 9, ..., 1936.

    A) 1      B) 21      C) 32      D) 44

    35.

    D

*The end of the contest* 🖝 **8**

**Visit our Web site at http://www.mathleague.com**

# Algebra Course 1 Solutions

## 2006-2007 through 2010-2011

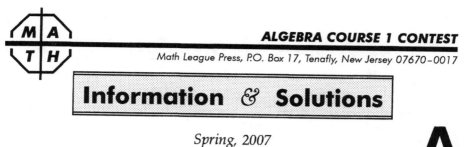
# Information & Solutions

*Spring, 2007*

## Contest Information

- **Solutions** Turn the page for detailed contest solutions (written in the question boxes) and letter answers (written in the *Answers* column to the right of each question).

- **Scores** Please remember that *this is a contest, not a test*—and there is no "passing" or "failing" score. Few students score as high as 30 points (75% correct). Students with half that, 15 points, *deserve commendation!*

- **Answers & Rating Scale** Turn to page 148 for the letter answers to each question and the rating scale for this contest.

1. $(3+6)^2 = 3^2 + 2\times3\times6 + 6^2 = 3^2+6^2 + 36$.

   A) 0          B) 9          C) 18          D) 36

2. Since $2(x+0) + 0(x+7) = 2x$, and $x = 2007$, $2x = 4014$.

   A) 2     B) 2007     C) 4007     D) 4014

3. I was able to compute that $(-2)(-3)(-4)(-5) =$ $(-3)(-4)(-2)(-5) = (-3)(-4)(10)$.

   A) –20     B) –10     C) 10     D) 20

4. If $\frac{a}{b} = -2$, then $a = -2b$, so $2b = -a$.

   A) $-a$     B) $-4a$     C) $a$     D) $4a$

5. If $x+1 = y$, then $x^2+2x+1 = (x+1)^2 = y^2$.

   A) $2y+1$          B) $y^2$          C) $y^2+1$          D) $y^2+2x$

6. If $x = 0$, then $x^2 = 0$, its least value.

   A) $-3 < x < -1$ B) $-2 < x < 1$   C) $0 < x < 1$     D) $1 < x < 2$

7. $10x^2-5x+(-3x)-(-2x^2) = 10x^2-5x-3x+2x^2 = 12x^2-8x$.

   A) $12x^2-8x$  B) $12x^2-2x$  C) $8x^2-8x$  D) $8x^2-2x$

8. Only 40% of the drummers passed me in the 1st hour, so 60% of the drummers passed me in the 2nd hour. Therefore, $60\% = 45, 20\% = 15$, and $100\% = 75$.

   A) 63     B) 70     C) 72     D) 75

9. Since $x+10$ is odd, $x$ must be odd; so $x+5 = $ odd$+5$ is even.

   A) odd          B) even          C) prime          D) positive

10. The roots of $(x-2)(x-3) = 0$ are 2, 3; their product is $2\times3 = 6$.

    A) 0          B) –5          C) –6          D) 6

11. The g.c.f. of $(x+1)(x+1)$ and $(x+1)(x+2)$ is $x+1$.

    A) 2          B) $x^2$          C) $x+1$          D) $x+2$

*Go on to the next page* ⟾ **A**

12. Slopes of ∥ lines are equal. When each is $\frac{1}{2}$, the sum is $\frac{1}{2}+\frac{1}{2}=1$.

    A) $\frac{1}{16}$     B) $\frac{1}{4}$     C) $\frac{1}{2}$     D) 1

**12. D**

13. The composite factors of $30 = 2\times3\times5$ are: $2\times3$, $2\times5$, $3\times5$, and $2\times3\times5$. The number of composite factors is 4.

    A) 3        B) 4        C) 6        D) 7

**13. B**

14. If $x < 0$, then $|x-2| = -(x-2) = -x+2 =$ A.

    A) $|x|+2$     B) $|x|-2$     C) $2-|x|$     D) $x+2$

**14. A**

15. $(x^2-4)(x^2-16)(x^2-20)(x^2-36) = 0$, so $x = \pm2, \pm4, \pm\sqrt{20}, \pm6$.

    A) 3              B) 4              C) 6              D) 8

**15. C**

16. Since (1,3) and (6,3) lie on a horizontal line, $m = 0$ and $mb = 0$.

    A) 6              B) 5              C) 3              D) 0

**16. D**

17. As shown, A, B, and D are all sums of squares of two integers.

    A) $17 = 1^2+4^2$   B) $18 = 3^2+3^2$   C) 19          D) $20 = 2^2+4^2$

**17. C**

18. If $x = 1000$, then 1% of $x = \frac{1}{100}(1000) = 10$, and 0.5% of $x = 5$.

    A) 0.5            B) 5            C) 50            D) 500

**18. B**

19. The sign's width is $(x-6)(x+1)$. Its length is $\frac{1}{(x-3)(x+1)}$. To get its area, multiply. Reduce the result to get $\frac{x-6}{x-3}$.

    A) 2                   B) $\frac{x-2}{x+1}$

    C) $\frac{x-6}{x-3}$         D) $\frac{5x+6}{2x+3}$

**19. C**

20. Since $\sqrt{2007} \approx 44.8$, $x = -44, -43, -42, \ldots, -2, -1$.

    A) 44            B) 45            C) 88            D) 89

**20. A**

21. C is *not* the product of two binomials with integer coefficients.

    A) $(x+y)(x+y)$   B) $(x-y)(x-y)$   C) $x^2-2xy-y^2$   D) $(x-y)(x+y)$

**21. C**

*Go on to the next page* ⏩ **A**

22. $\sqrt{x^{64}} \div \sqrt{x^4} = x^{32} \div x^2 = x^{30}$.

   A) $x^4$     B) $x^6$     C) $x^{16}$     D) $x^{30}$

22.

D

23. The number of times that I danced with a star is $(x+3)^2-(x-3)^2 =$ $(x^2+6x+9)-(x^2-6x+9) = 12x$.

   A) 0   B) $3x$   C) $12x$   D) 18

23.

C

24. $x^2+x^2+x^2+x^2 = 4x^2 = x^4$, so $x = 2$ & $x^4+x^4+x^4+x^4 = 4x^4 = 64 = 8x^3$.

   A) $x^8$     B) $4x^6$     C) $6x^4$     D) $8x^3$

24.

D

25. $(x+y)^3 = 1x^3+3x^2y+3xy^2+1y^3$. The coefficients' sum is $1+3+3+1$.

   A) 8     B) 6     C) 4     D) 3

25.

A

26. The square factors are 1, $2^2$, $2^4$, $5^2$, $5^4$, $2^2 \times 5^2$, $2^2 \times 5^4$, $2^4 \times 5^2$, $2^4 \times 5^4$.

   A) 4     B) 9     C) 16     D) 25

26.

B

27. B crosses $x$-axis at $y = 0$. Only $0 = x^2-64$ has unequal solutions.

   A) $y = (x+8)(x-8)$         B) $y = x^2$
   C) $y = (x+5)^2$           D) $y = (x-12)^2$

27.

A

28. Since line $\ell$ has slope $\dfrac{1}{1+\frac{1}{x}} = \dfrac{x}{x+1}$, the slope of any line perpendicular to line $\ell$ is the negative reciprocal of $\dfrac{x}{x+1}$, and that's choice C.

   A) $-x$     B) $-\dfrac{x}{x+1}$     C) $-\dfrac{x+1}{x}$     D) $-\dfrac{1}{x+1}$

28.

C

29. $10^3+10^4=1000+10000=11000$. Likewise, $10^{2006}+10^{2007}$ has 2008 digits.

   A) 2007     B) 2008     C) 4013     D) 4015

29.

B

30. $(x+y)^2 = 36 = x^2+2xy+y^2$. Since $xy = 4$, $x^2+y^2 = 36-8 = 28$.

   A) 20     B) 28     C) 32     D) 36

30.

B

*The end of the contest* ☜ **A**

**Visit our Web site at http://www.mathleague.com**

120

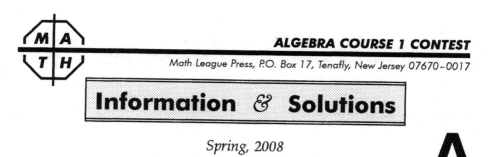
# Information & Solutions

*Spring, 2008*

## Contest Information

**A**

- **Solutions** Turn the page for detailed contest solutions (written in the question boxes) and letter answers (written in the *Answers* column to the right of each question).

- **Scores** Please remember that *this is a contest, not a test*—and there is no "passing" or "failing" score. Few students score as high as 30 points (75% correct). Students with half that, 15 points, *deserve commendation!*

- **Answers & Rating Scale** Turn to page 149 for the letter answers to each question and the rating scale for this contest.

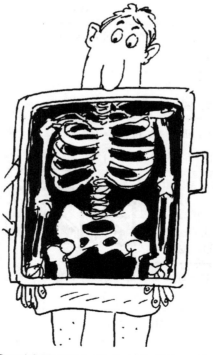

1. $(2 + 0 + 0 + 8)^0 = 10^0 = 1.$

    A) 0      B) 1      C) 4      D) 9

    1. B

2. If $x^2 = 10$, then $(x+1)(x-1) = x^2 - 1 = 10 - 1 = 9.$

    A) 99      B) 11      C) 9      D) −1

    2. C

3. The 36 houses painted in 2007 were 50% more than Joe painted in 2006. Since 12 is 50% of 24, and $24+12 = 36$, Joe painted 24 houses in 2006.

    A) 12      B) 18      C) 24      D) 27

    3. C

4. $x^2-1 = (x+1)(x-1)$ is divisible by $x+1$, so the fraction in A reduces to $x-1$.

    A) $\dfrac{x^2-1}{x+1}$      B) $\dfrac{x^2-2}{x+2}$

    C) $\dfrac{x^2-3}{x+3}$      D) $\dfrac{x^2-4}{x+4}$

    4. A

5. The 2009 factors of $2^{2008}$ are $1, 2^1, 2^2, \ldots, 2^{2007},$ and $2^{2008}.$

    A) 1      B) 2007      C) 2008      D) 2009

    5. D

6. Since $100(x+y) = (x+y)(x+y)$, it follows that $100 = x+y.$

    A) 10      B) 100      C) $100+100$      D) $100^2$

    6. B

7. $(x^2+2x+1) - (x^2-2x+1) = x^2+2x+1-x^2+2x-1 = 4x.$

    A) 0      B) $-2x^2$      C) $2x$      D) $4x$

    7. D

8. Since $a \geq b > 0$ and $ab = 64$, $(a,b) = (64,1), (32,2), (16,4),$ or $(8,8).$

    A) 7      B) 6      C) 4      D) 3

    8. C

9. The # between 50 & 150 whose $\sqrt{\ }$ is prime is 121. The sum of its digits is 4.

    A) 2      B) 4

    C) 11      D) 121

    9. B

10. $x^2+5x-6 = (x+6)(x-1)$ is divisible by $x+6$ and $x-1.$

    A) $x-6$      B) $x-3$      C) $x-2$      D) $x-1$

    10. D

11. Since $|x-y| > x-y$, it follows that $x-y < 0$, from which $y > x.$

    A) $y > x$      B) $x > y$      C) $y < 0$      D) $x > 0$

    11. A

*Go on to the next page* ⫸ **A**

| | |
|---|---|
| 12. The only 16 consecutive integers whose sum is 8 are the integers –7, –6, . . . , 7, 8.<br>A) –6    B) –7    C) –8    D) –16 | 12.<br>B |
| 13. The $y$-intercept of $y = x+1$ is 1, so let $x = 1$ and $y = 0$. That works in choice A.<br>A) $0 = 1-1$    B) $0 = 1$<br>C) $0 = 1+1$    D) $0 = 1^2$ | 13.<br>A |
| 14. Since $(x-1)(x+1)(x-2)(x+2) = (x^2-1)(x^2-4) = x^4-5x^2+4$, the result, with like terms combined, has exactly 3 terms.<br>A) 3    B) 6    C) 9    D) 12 | 14.<br>A |
| 15. $x[x(x^2)^2]^2 = x[x(x^4)]^2 = x[x^5]^2 = x[x^{10}] = x^{11}$.<br>A) $x^8$    B) $x^9$    C) $x^{10}$    D) $x^{11}$ | 15.<br>D |
| 16. $x^2+bx+12 = (x\pm1)(x\pm12)$ or $(x\pm2)(x\pm6)$ or $(x\pm3)(x\pm4)$.<br>A) 6    B) 5    C) 4    D) 3 | 16.<br>A |
| 17. The line $y = -4x$ goes through QII, the origin, and QIV.<br>A) $y = 2x+4$   B) $y = -2x+4$   C) $y = 4x$   D) $y = -4x$ | 17.<br>D |
| 18. Since $\frac{1}{4x^4-4}$ is undefined if $4x^4-4 = 0$, it's undefined if $x = \pm1$.<br>A) 0    B) $\pm1$    C) $\pm2$    D) $\pm4$ | 18.<br>B |
| 19. $999\,999^{6x} \div 999\,999^{3x} = 999\,999^{6x-3x} = 999\,999^{3x}$.<br>A) $999\,999^2$   B) $999\,999^{2x}$   C) $999\,999^{3x}$   D) $999\,999^3$ | 19.<br>C |
| 20. Write each in the form $y = mx+b$. Choice D has slope 2.<br>A) $y = -x+2$   B) $y = x-2$   C) $y = -2x+2$   D) $y = 2x-2$ | 20.<br>D |
| 21. Since $\frac{2.2}{3.3+4.4} \times \frac{10}{10} = \frac{22}{33+44}$, our score and theirs would have been the same if they also had a score of $\frac{22}{33+44}$ + 1.1.<br>A) 0.011    B) 0.11<br>C) 1.1    D) 11 | 21.<br>C |
| 22. If $x < 0$, then $x^2 > 0$ and $x^3 < 0$, so $x^2 > x^3$.<br>A) $x > x^2$   B) $x > x^3$   C) $x^2 > x^3$   D) $x^3 > x^2$ | 22.<br>C |

*Go on to the next page* ⟹ **A**

23. The average number that the pigs wore was $(1+p)/2$. Multiply this average by the number of pigs to get the total, which is $p(1+p)/2$.

    A) $\frac{1}{2}p(1+p)$    B) $p(1+p)$

    C) $\frac{1}{2}(1+p)$    D) $\frac{1}{2}p(p-1)$

24. $2\sqrt{8} = 4\sqrt{2}$. Adding this to $8\sqrt{2}$, the sum is $12\sqrt{2} = \sqrt{288}$.

    A) $\sqrt{256}$    B) $\sqrt{288}$    C) $\sqrt{384}$    D) $\sqrt{512}$

25. Only integer pairs $(-1,-1)$, $(-1,1)$, $(1,-1)$, $(1,1)$ satisfy $x^2+y^2 = 2$.

    A) 2    B) 4    C) 6    D) 8

26. If $x$ is an even integer $> 0$, then $(-x)^x = x^x$, so $-x^x + x^x = 0$.

    A) positive    B) negative    C) zero    D) undefined

27. $x^2+3\pi x+2\pi^2 = (x+\pi)(x+2\pi) = (2)(2+\pi) = 4+2\pi$.

    A) $2+2\pi$    B) $4+\pi$    C) $2+\pi$    D) $4+2\pi$

28. The sides of the x-ray screen have lengths $5x$ and $4x$, so the screen's area is $20x^2 = 20 \times$ (a perfect square). Divide each choice by 20. Only $8000 \div 20 = 400 = 20^2$ is the square of an integer.

    A) 200    B) 600    C) 4000    D) 8000

29. Avg. speed = (total distance) ÷ (total time), so $\frac{1}{k} = 2\div(\frac{1}{k} + \frac{1}{r})$. Cross-multiply to get $\frac{1}{k} + \frac{1}{r} = 2k$. Now, $\frac{1}{r} = 2k - \frac{1}{k}$; so $r = \frac{k}{2k^2-1}$.

    A) $\frac{k^2-2}{k}$    B) $\frac{2-k^2}{k}$    C) $\frac{k}{2k^2-1}$    D) $\frac{k^2+1}{2k}$

30. The 2008th term of the sequence is $\frac{\sqrt{2^{2008}}}{2^{2008}} = \frac{2^{1004}}{2^{2008}} = \frac{1}{2^{1004}}$.

    A) $\frac{1}{2^{1004}}$    B) $\frac{1}{1004}$    C) $\frac{1}{2^{2008}}$    D) $\frac{1}{2008}$

*The end of the contest* ✍ **A**

# Information & Solutions

*Spring, 2009*

## Contest Information

- **Solutions** Turn the page for detailed contest solutions (written in the question boxes) and letter answers (written in the *Answer Column* to the right of each question).

- **Scores** Please remember that *this is a contest, and not a test* — there is no "passing" or "failing" score. Few students score as high as 24 points (80% correct); students with half that, 12 points, *deserve commendation!*

- **Answers and Rating Scales** Turn to page 150 for the letter answers to each question and the rating scale for this contest.

1.  If $m = 2$, $a = 3$, and $t = 5$, then $2 + 3 + 5 + h = 22$.
    So $10 + h = 22$, and $h = 12$.

    A) 4          B) 7          C) 9          D) 12

2.  21 centuries = $21 \times 100$ years = 2100 years.
    2100 years = $210 \times (10$ years$)$ = 210 decades.

    A) 12          B) 21          C) 210          D) 2100

3.  $(x + 2x + 4x + 6x) + (1 + 3 + 5 + 7) = 13x + 16$.

    A) $12x + 16$          B) $13x + 16$
    C) $13x + 12$          D) $16x + 13$

4.  For the square of a number to be less than the number itself, the
    number must be positive and less than 1. The only choice that is
    both positive and less than 1 is choice A.

    A) $\dfrac{1}{3}$          B) $-\dfrac{1}{3}$          C) 3          D) –3

5.  $(x + 1)(x - 1) - (x + 2)(x - 2) = (x^2 - 1) - (x^2 - 4) = x^2 - 1 - x^2 + 4 = 3$.

    A) –5          B) 3          C) $2x^2 - 5$          D) $2x^2 + 3$

6.  If $x^2 = 3$, then $x^4 = (x^2)^2 = (3)^2 = 9$. Thus, $x^4 - 3 = 6$.

    A) 0          B) 3          C) 6          D) 9

7.  $x + 200\%$ of $x = x + 2x = 3x$.

    A) $x + 200$          B) $x + 300$          C) $2x$          D) $3x$

8.  If $x = y$, then $(x - y) = 0$. $(x + y)(x - y) = (x + y)(0) = 0$.

    A) 0          B) $x^2 - 2xy + y^2$          C) $x^2 + 2xy - y^2$          D) $x^2 + y^2$

9.  $cd^2 = c \times d^2 = c \div 1/d^2$.

    A) $\dfrac{1}{d}$          B) $\dfrac{1}{d^2}$          C) $\dfrac{d^2}{c}$          D) $\dfrac{c}{d^2}$

10. Since $10p$ is divisible by 10 and $p$, $10p$ cannot be
    a prime number.

    A) $p + 10$          B) $p - 10$          C) $10p$          D) $p + 1000$

11. If the greatest of the integers is $x$, then
    $x + (x - 1) + (x - 2) + (x - 3) + (x - 4) = 10055$.
    Therefore , $5x - 10 = 10055$, $5x = 10065$,
    and $x = 2013$.

    A) 2009          B) 2010          C) 2012          D) 2013

*Go on to the next page* ))))**➤** **A**

12. In $y$ min, Ben runs $xy$ m. Since 1 m = 1/1000 km, $xy$ m = $xy$/1000 km.

    A) $\dfrac{x}{1000y}$    B) $\dfrac{y}{1000x}$    C) $\dfrac{xy}{1000}$    D) $\dfrac{1000}{xy}$

    12.

    C

13. $5^{777} = 5^{555} \times 5^{222}$, so both are divisible by $5^{555}$.

    A) 5    B) $5^{111}$    C) $5^{555}$    D) $5^{777}$

    13.

    C

14. For any $y > 1$, the larger the exponent of $y$, the larger the value. The largest exponent is in A.

    A) $y^{2008}$    B) $\sqrt{y^{4008}}$    C) $(y^{200})^8$    D) $y^2 y^{1008}$

    14.

    A

15. The intercepts are $(x,0)$ and $(0,y)$; the slope will be $(y - 0) \div (0 - x) =$ $y \div (-x) = -(y \div x)$. Both $y$ and $x$ are positive, so $-(y \div x)$ is negative.

    A) positive    B) negative    C) 0    D) undefined

    15.

    B

16. $2^{100} \times (-2)^{101} \div 2^{202} = 2^{100} \times (-2^{101}) \div 2^{202} = -2^{201} \div 2^{202} = -1 \div 2^1 = -1/2$.

    A) 2    B) –2    C) $\dfrac{1}{2}$    D) $-\dfrac{1}{2}$

    16.

    D

17. If $(x - 11)^2 = 1$, then $x - 11 = 1$ or $x - 11 = -1$. So, $x = 12$ or $x = 10$.

    A) –2    B) 1    C) 2    D) 22

    17.

    D

18. If $x = 25$, then $x^3 = 25^3 = (5^2)^3 = (5^3)^2 = 125^2$.

    A) 20    B) 25    C) 30    D) 35

    18.

    B

19. Perpendicular lines have negative reciprocal slopes; if the slope of $\ell$ is 2, the perpendicular slope is –1/2, and the sum 3/2.

    A) $\dfrac{3}{2}$    B) $\dfrac{7}{3}$    C) $\dfrac{9}{5}$    D) $\dfrac{8}{7}$

    19.

    A

20. Charlie didn't make such a good deal! $x = (-6) + (-5) + (-4) + (-3) + (-2) + (-1) + 0 + 1 + 2 + 3 + 4 + 5 + 6 = 0$. Since Charlie's teacher agrees to pay Charlie $5 + x$, his pay is $5 regardless of how many hours he works.

    A) \$5    B) \$26    C) \$131    D) \$156

    20.

    A

21. Plot (2,6) and (14,1). The $y$-coordinate of the third vertex is 6, so move along the line $y = 6$ to find a point that forms a right angle when connected to the two given vertices.

    A) 1    B) 2    C) 6    D) 14

    21.

    D

*Go on to the next page* )))➡ **A**

| | Answers |
|---|---|
| 22. The sum of three or more consecutive terms is never as large as the first term and never as small as the sum of the first two terms. <br><br> A) $-\dfrac{1}{3} < x < -\dfrac{1}{12}$  B) $-\dfrac{1}{12} < x < 0$  C) $0 < x < \dfrac{1}{12}$  D) $\dfrac{1}{12} < x < \dfrac{1}{3}$ | 22. <br><br> D |
| 23. If $|x + y| > x + y$, then $x + y < 0$. Either $x < 0$ and $y < 0$, or the variable with the greater absolute value is negative. <br><br> A) $x < 0$   B) $x > 0$   C) $y < 0$   D) $y > 0$ | 23. <br><br> A |
| 24. Use algebraic division or note that $x^3 - 2x + 1 = (x^3 - x^2) + (x^2 - 2x + 1) = -x^2(1 - x) + (1 - x)(1 - x)$. <br><br> A) $-x^2 - x + 1$ $\qquad$ B) $-x^2 + x - 1$ <br> C) $x^2 - x - 1$ $\qquad$ D) $x^2 + x - 1$ | 24. <br><br> A |
| 25. $63 = 3^2 \times 7^1$ needs one 3 and two 7s to be a cube. $3^1 \times 7^2 = 147$. <br><br> A) 7 $\qquad$ B) 83 $\qquad$ C) 147 $\qquad$ D) 3969 | 25. <br><br> C |
| 26. The area of a circle of radius $r$ is $\pi r^2$. The side-length of the square is $2r$, so the area of the square is $(2r)^2 = 4r^2$. The ratio is thus $\pi r^2/4r^2 = \pi/4$. <br><br> A) $\dfrac{2}{r}$ $\qquad$ B) $\dfrac{\pi}{4}$ $\qquad$ C) $\dfrac{\pi r}{4}$ $\qquad$ D) $\dfrac{r}{4}$ | 26. <br><br> B |
| 27. Factoring, $y = (x - 2)(x - 4)[(x - 6) + (x - 8) + (x - 10)] = (x - 2)(x - 4)(3x - 24) = 3(x - 2)(x - 4)(x - 8)$. When $y = 0$, $x = 2$, 4, or 8. These are the only points at which the graph crosses the $x$-axis. <br><br> A) 2 $\qquad$ B) 4 $\qquad$ C) 6 $\qquad$ D) 8 | 27. <br><br> C |
| 28. The students' heights total 150 cm $\times$ 56 = 8400 cm. If the boys' heights total 165 cm $\times$ 21 = 3465 cm, then the girls' heights must total 8400 − 3465 = 4935 cm. The average girl's height is 4935 cm $\div$ 35 = 141 cm. <br><br> A) 135 $\qquad$ B) 141 $\qquad$ C) 151 $\qquad$ D) 155 | 28. <br><br> B |
| 29. $5^n + 5^n + 5^n + 5^n + 5^n = 5(5^n) = 5^{n+1}$; $n + 1 = 50$; $n = 49$. <br><br> A) 10 $\qquad$ B) 21 $\qquad$ C) 38 $\qquad$ D) 49 | 29. <br><br> D |
| 30. $2(2x - 5y) + 3(3x + 4y) = 13x + 2y$; $2(11) + 3(7) = 43$. <br><br> A) 4 $\qquad$ B) 18 $\qquad$ C) 43 $\qquad$ D) 62 | 30. <br><br> C |

*The end of the contest* ✍ **A**

**Visit our Web site at http://www.mathleague.com**

# Information & Solutions

*Spring, 2010*

## Contest Information

■ **Solutions**   Turn the page for detailed contest solutions (written in the question boxes) and letter answers (written in the *Answer Column* to the right of each question).

■ **Scores**   Please remember that *this is a contest, and not a test*—there is no "passing" or "failing" score. Few students score as high as 24 points (80% correct); students with half that, 12 points, *deserve commendation!*

■ **Answers and Rating Scales**   Turn to page 151 for the letter answers to each question and the rating scale for this contest.

1.  If $b = 1$, $e = 2b = 2$, $a = 3e = 6$, and $r = 4a = 24$,
    then $b + e + a + r = 1 + 2 + 6 + 24 = 33$.

    A) 4          B) 10          C) 24          D) 33

2.  When $x = 1$, each power of $x$ is also 1. Therefore,
    the values of the choices are 0, –1, –2, and –3.

    A) $x - 1$     B) $x^2 - 2$     C) $x^3 - 3$     D) $x^4 - 4$

3.  $(-4)^2(-3)^0(-2)^1(-1)^0 = (16)(1)(-2)(1) = -32$.

    A) –32     B) 0          C) 16          D) 32

4.  $x^4 - 16 = (x^2 - 4)(x^2 + 4) = (x - 2)(x + 2)(x^2 + 4)$.

    A) $x + 4$          B) $x - 4$          C) $x + 2$          D) $x - 1$

5.  $x^{-2010}$ is the reciprocal of $x^{2010}$. For positive values of $x^{-2010}$, $x^{-2010}$ is
    greatest when $x^{2010}$ is least, and $x^{2010}$ is least when $x = 100$.

    A) 100          B) 200          C) 300          D) 400

6.  If $300x = 450 - 300y$, then $300x + 300y = 450$ and $x + y = 1.5$.

    A) –3          B) 1.5          C) 15          D) 30

7.  $(2x^4 + 4x^2) + (3x^4 - 5x^2) - (4x^4 - 6x^2) = (2x^4 + 3x^4 - 4x^4) + (4x^2 - 5x^2 + 6x^2)$.

    A) $x^4 - 7x^2$     B) $x^4 + 5x^2$     C) $x^4 - 3x^2$     D) $x^4 + x^2$

8.  If $x > 0$, then the additive inverse of $x$ divided by the reciprocal of $x$
    equals $-x$ divided by $\frac{1}{x} = (-x)(x) = -x^2$.

    A) the square of $x$          B) the reciprocal of the square of $x$
    C) the square root of $x$     D) the additive inverse of the square of $x$

9.  $(a^4)^3 = a^{12}$ and $(a^{12})^2 = a^{24}$.

    A) $a^9$     B) $a^{24}$     C) $a^{36}$     D) $a^{64}$

10. If $g - c = 2$ and $(g - c)(g + c) = 20$, then
    $2(g + c) = 20$ and $g + c = 10$. Thus, $g = 6$ and $c = 4$.

    A) 6          B) 8          C) 10          D) 12

11. $\frac{x}{100} \times \frac{y}{100} \times 100\,000 = 10xy$.

    A) $x + y$     B) $xy$     C) $\frac{x + y}{10}$     D) $10xy$

12. $\sqrt{2} + \sqrt{4} + \sqrt{8} + \sqrt{16} = \sqrt{2} + 2 + 2\sqrt{2} + 4 = 6 + 3\sqrt{2}$.

    A) $\sqrt{30}$     B) $6 + 3\sqrt{2}$     C) $10\sqrt{2}$     D) $3 + 4\sqrt{2}$

*Go on to the next page* ))))➡ **A**

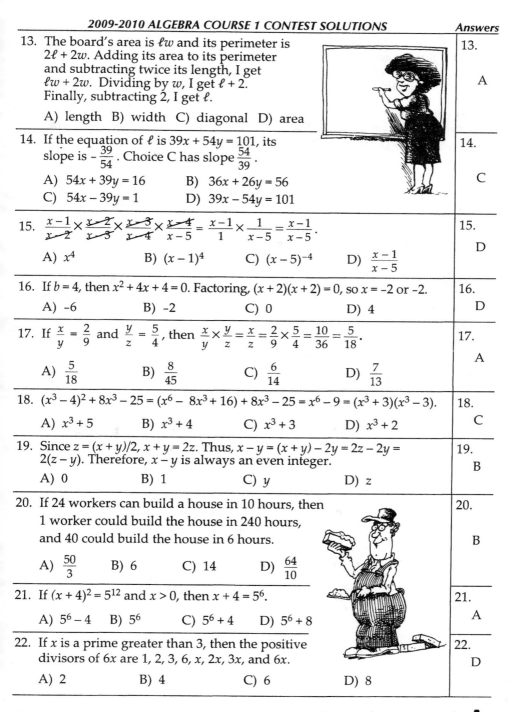

13. The board's area is $\ell w$ and its perimeter is $2\ell + 2w$. Adding its area to its perimeter and subtracting twice its length, I get $\ell w + 2w$. Dividing by $w$, I get $\ell + 2$. Finally, subtracting 2, I get $\ell$.

  A) length   B) width   C) diagonal   D) area

**13. A**

14. If the equation of $\ell$ is $39x + 54y = 101$, its slope is $-\frac{39}{54}$. Choice C has slope $\frac{54}{39}$.

  A) $54x + 39y = 16$       B) $36x + 26y = 56$
  C) $54x - 39y = 1$       D) $39x - 54y = 101$

**14. C**

15. $\frac{x-1}{x-2} \times \frac{x-2}{x-3} \times \frac{x-3}{x-4} \times \frac{x-4}{x-5} = \frac{x-1}{1} \times \frac{1}{x-5} = \frac{x-1}{x-5}$.

  A) $x^4$       B) $(x-1)^4$       C) $(x-5)^{-4}$       D) $\frac{x-1}{x-5}$

**15. D**

16. If $b = 4$, then $x^2 + 4x + 4 = 0$. Factoring, $(x+2)(x+2) = 0$, so $x = -2$ or $-2$.

  A) $-6$       B) $-2$       C) $0$       D) $4$

**16. D**

17. If $\frac{x}{y} = \frac{2}{9}$ and $\frac{y}{z} = \frac{5}{4}$, then $\frac{x}{y} \times \frac{y}{z} = \frac{x}{z} = \frac{2}{9} \times \frac{5}{4} = \frac{10}{36} = \frac{5}{18}$.

  A) $\frac{5}{18}$       B) $\frac{8}{45}$       C) $\frac{6}{14}$       D) $\frac{7}{13}$

**17. A**

18. $(x^3 - 4)^2 + 8x^3 - 25 = (x^6 - 8x^3 + 16) + 8x^3 - 25 = x^6 - 9 = (x^3 + 3)(x^3 - 3)$.

  A) $x^3 + 5$       B) $x^3 + 4$       C) $x^3 + 3$       D) $x^3 + 2$

**18. C**

19. Since $z = (x+y)/2$, $x + y = 2z$. Thus, $x - y = (x+y) - 2y = 2z - 2y = 2(z-y)$. Therefore, $x - y$ is always an even integer.

  A) $0$       B) $1$       C) $y$       D) $z$

**19. B**

20. If 24 workers can build a house in 10 hours, then 1 worker could build the house in 240 hours, and 40 could build the house in 6 hours.

  A) $\frac{50}{3}$    B) $6$    C) $14$    D) $\frac{64}{10}$

**20. B**

21. If $(x+4)^2 = 5^{12}$ and $x > 0$, then $x + 4 = 5^6$.

  A) $5^6 - 4$    B) $5^6$    C) $5^6 + 4$    D) $5^6 + 8$

**21. A**

22. If $x$ is a prime greater than 3, then the positive divisors of $6x$ are 1, 2, 3, 6, $x$, $2x$, $3x$, and $6x$.

  A) $2$       B) $4$       C) $6$       D) $8$

**22. D**

Go on to the next page ⟫⟫➡ **A**

23. Use the Pythagorean Theorem. In 1 hour at these rates, Fred would be $c$ km from his starting point where $c^2 = 1^2 + (3/4)^2$. Simplifying, $c^2 = 25/16$ and $c = 5/4$. In 1 min. Fred would travel $(5/4 \div 60)$ km.

A) $\dfrac{1}{100}$    B) $\dfrac{1}{70}$    C) $\dfrac{1}{48}$    D) $\dfrac{1}{20}$

23.

C

24. Since $100^{100} = (100^{50})^2$, it follows that $\sqrt{100^{100}} = 100^{50}$.

A) $10^{10}$    B) $10^{50}$    C) $100^{10}$    D) $100^{50}$

24.
D

25. $(x + 1)^4 = (x + 1)^2(x + 1)^2 = (x^2+2x+1)^2 = x^4 + 4x^3 + 6x^2 + 4x + 1$. The desired sum is $4 + 4 = 8$.

A) 4    B) 6    C) 8    D) 10

25.

C

26. $(x + 4)^2 = (y + 5)^2$, so $x^2 + 8x + 16 = y + 10y + 25$. Thus, $x^2 + 8x = y^2 + 10y + 9$.

A) $y^2 + 10y + 9$    B) $y^2 + 9y + 10$    C) $y^2 + 10y$    D) $y^2 + 9y$

26.
A

27. $12^{1200} = (3^{1200})(4^{1200}) = (3^{1200})(2^2)^{1200} = (3^{1200})(2^{2400})$, so $x = 2400$.

A) 600    B) 1200    C) 2400    D) 3600

27.

C

28. Let $3f$ = # of grandfathers and $4m$ = # of grandmothers. If $\dfrac{2}{3}$ of the grandfathers and $\dfrac{1}{2}$ of the grandmothers bought coats, then $2f$ grandfathers bought coats and $2m$ grandmothers bought coats. We know that $3f + 4m = 100$ and $2f = 3m + 10$. Solving, $f = 20$ and $m = 10$. Thus, # of grandfathers who bought coats = $2f = 40$.

A) 35    B) 40    C) 55    D) 60

28.

B

29. If $r = (2 + 4 + 6 + \ldots + 2010)$ and $s = (1 + 3 + 5 + \ldots + 2009)$, then $r - s = (2 - 1) + (4 - 3) + (6 - 5) + \ldots + (2010 - 2009) = 1 + 1 + 1 + \ldots + 1 = 1005$.

A) 1005    B) 1006    C) 2010    D) 2011

29.

A

30. Clearing fractions, $(y^2 - x^2) + (y - x) = 0$. Factoring, $(y - x)(y + x) + (y - x) = 0$. Dividing by $(y - x)$, which is not 0, $y + x + 1 = 0$, so $x + y = -1$.

A) –2    B) –1    C) 0    D) 1

30.
B

*The end of the contest*  **A**

# Information & Solutions

*Spring, 2011*

## Contest Information

- **Solutions**   Turn the page for detailed contest solutions (written in the question boxes) and letter answers (written in the *Answer Column* to the right of each question).

- **Scores**   Please remember that *this is a contest, and not a test*—there is no "passing" or "failing" score. Few students score as high as 24 points (80% correct); students with half that, 12 points, *deserve commendation!*

- **Answers and Rating Scales**   Turn to page 152 for the letter answers to each question and the rating scale for this contest.

1. If $xy = 2011^2$, then $(-x)(-y) = (-1)(x)(-1)(y) = xy = 2011^2$.

   A) $-2011^2$     B) $2011^{-2}$     C) $-2011^{-2}$     D) $2011^2$

2. One week is 7 days, so $w$ weeks = $7w$ days.

   A) $\dfrac{w}{7}$     B) $w + 7$     C) $7w$     D) $w^7$

3. Since $x^2 - 4x - 12 = (x + 2)(x - 6)$ and
   $(x + 2)(x - d) = x^2 - 4x - 12$, then $x - 6 = x - d$
   and $d = 6$.

   A) $-8$     B) $-6$     C) $6$     D) $8$

4. $\sqrt{4x} + \sqrt{9x} + \sqrt{25x} = 2\sqrt{x} + 3\sqrt{x} + 5\sqrt{x} = 10\sqrt{x}$.

   A) $10\sqrt{x}$     B) $10 + \sqrt{x}$     C) $10 + 3\sqrt{x}$     D) $30\sqrt{x}$

5. $(x - 2x) + (3x - 4x) + (5x - 6x) + (7x - 8x) + (9x - 10x) = -5x$.

   A) $-10x$     B) $-5x$     C) $-2x$     D) $0$

6. The sum of five consecutive integers is 165. The middle one is the
   average = $165 \div 5 = 33$. The integers are 31, 32, 33, 34, and 35.

   A) $31$     B) $33$     C) $35$     D) $37$

7. $\dfrac{x}{z} \div \dfrac{z}{x} = \dfrac{x}{z} \times \dfrac{x}{z} = \dfrac{x^2}{z^2}$.

   A) $1$     B) $\dfrac{x^2}{z^2}$     C) $\dfrac{z^2}{x^2}$     D) $x^2 z^2$

8. Lois and Clark are above the ground. Therefore,
   $h > 0$, $h + 2 = 28$, and $h = 26$.

   A) $26$     B) $52$     C) $26^2$     D) $52^2$

9. $(x+5)^2 - (x-5)^2 = (x^2+10x+25) - (x^2-10x+25) = 20x$.

   A) $0$     B) $50$     C) $10x$     D) $20x$

10. If $p$ is a prime between 1000 and 2000, then $p$ is
    odd. Thus, $p + 567$ is even and is not a prime.

    A) $p - 10$     B) $p + 300$     C) $p + 456$     D) $p + 567$

11. If $2m+3s = 86$ and $3m+4s = 120$, then multiply 1st
    equation by 4 and 2nd by 3 to get $8m+12s = 344$ (1)
    and $9m+12s = 360$ (2). Subtracting (1) from (2), $m = 16$.

    A) $15$     B) $16$     C) $17$     D) $18$

*Go on to the next page* )))➡ **A**

12. If $n^2 + 5n = 24$ and $n^2 - 4n = -3$, then subtract to get $9n = 24 - (-3) = 27$.

   A) -9       B) 9       C) 27       D) 81

12.
C

13. If $a < 0 < b$, then $1/a$ is negative, $1/b$ is positive, and choice A is false.

   A) $\dfrac{1}{a} > \dfrac{1}{b}$     B) $a^2 > b^2$     C) $\dfrac{a}{b} > \dfrac{b}{a}$     D) $\dfrac{a^2}{b^2} > \dfrac{b^2}{a^2}$

13.
A

14. Since $(x^3 - 4x^2 + 4x - 3) \div (x - 3) = x^2 - x + 1$, it is divisible by $x - 3$. Of the polynomials listed, only $x^3 - 4x^2 + 4x - 3$ is divisible by $x - 3$.

   A) $x^2 + 9$       B) $x^2 + 8x + 15$

   C) $x^3 - 4x^2 + 4x - 3$     D) $x^3 - 2x^2 + 3x - 4$

14.
C

15. If $(x + 2010)(x+1) = 0$, then $x = -2010$ or $-1$.

   A) -2011    B) -2010    C) 2010    D) 2011

15.
A

16. The slope of $y = x$ is 1. Choice C has slope -1, the negative reciprocal.

   A) $3x + 4y = 5$    B) $5x - 5y = 13$    C) $6x + 6y = 19$    D) $8x - 7y = 31$

16.
C

17. $y^2 = (x - 5)^2 = x^2 - 10x + 25$. Therefore, $x^2 - 10x + 20 = y^2 - 5$.

   A) $y^2$       B) $y^2 - 5$       C) $y^2 - 10y + 25$   D) $y^2 + 10y + 15$

17.
B

18. If $(a + b)^2 = a^2 + 2ab + b^2 = 7^2$ and $a^2 + b^2 = 49$, then $2ab = 0$ and $ab = 0$.

   A) 42       B) 36       C) 7       D) 0

18.
D

19. If $x = 2z - y$, then $x + y = 2z$ and $z = (x + y)/2$.

   A) $2x - y$     B) $\dfrac{x+y}{2}$     C) $\dfrac{y-x}{2}$     D) $\dfrac{x}{2} + y$

19.
B

20. Let $5x$ and $3x$ be the initial numbers of apple cores and bottles. With 3 more bottles, $5x/(3x +3) = 3/2$. Simplifying, $10x = 9x + 9$. Therefore, $x = 9$, and $5x = 45$.

   A) 15     B) 25     C) 35     D) 45

20.
D

21. $2x \div 0.2x = 10 = 1000\%$.

   A) 1000%   B) 900%    C) 100%    D) 10%

21.
A

22. $|2x| + |-3x| = |2x| + |3x| = |5x| = 5|x|$.

   A) $|-x|$    B) $-|x|$    C) $-|5x|$    D) $5|x|$

22.
D

*Go on to the next page* )))➤ **A**

| | |
|---|---|
| 23. Set each distinct factor of the equation equal to 0:<br>$x^2 - 1 = 0$, $x^2 - 2 = 0$, $x^2 - 3 = 0$, ..., $x^2 - 20 = 0$.<br>The roots are $\pm 1$, $\pm\sqrt{2}$, ..., $\pm\sqrt{20}$. The integral<br>roots are $\pm 1$, $\pm 2$, $\pm 3$, $\pm 4$. There are 8 integers in all.<br><br>A) 4        B) 8        C) 20        D) 40 | 23.<br><br>B |
| 24. If $2011^{x^2+10x+21} = 1$, $x^2 + 10x + 21 = (x+7)(x+3) = 0$.<br>Thus, $x = -7$ or $-3$. The product of $-7$ and $-3$ is 21.<br><br>A) -21        B) -10        C) 10        D) 21 | 24.<br><br>D |
| 25. Since $99n = 3^2 \times 11n$ is a cube, the least such cube is $3^3 \times 11^3$;<br>so $n = 3 \times 11^2 = 363$. Finally, $3 + 6 + 3 = 12$.<br><br>A) 27        B) 18        C) 12        D) 9 | 25.<br><br>C |
| 26. The area is 480, so $w(w + 14) = 480$. Equivalently, $w^2 + 14w - 480 =$<br>$(w + 30)(w - 16) = 0$. Finally, $w = 16$, and the perimeter is $2(16+30) = 92$.<br><br>A) 88        B) 92        C) 116        D) 172 | 26.<br><br>B |
| 27. Divide $18x + 27y - 36 = 0$ by 9, then multiply by 2 to get $4x + 6y - 8 = 0$.<br><br>A) 0        B) 12        C) 24        D) 36 | 27.<br><br>A |
| 28. The graph of $y = |2x-9| - |2x+9|$ consists of 3 lines. If $x \geq 9/2$, the graph<br>is the line $y = -18$. If $x \leq -9/2$, the graph is the line<br>$y = 18$. If $-9/2 < x < 9/2$, the graph is the line $y = -4x$.<br>Since $y$ ranges from $-18$ to 18, $y$ can equal 15, but $y$<br>cannot equal any of the other values listed below.<br><br>A) 15        B) 19        C) 20        D) 22 | 28.<br><br>A |
| 29. The ones digits of powers of 123 are 3, 9, 7, 1, 3, ....<br>Every 4th power ends in 1; so $r$ ends in 1, as does $s$.<br><br>A) 1        B) 3        C) 7        D) 9 | 29.<br><br>A |
| 30. The least common multiple of all the integers from<br>1 through 30 is $2^4 \times 3^3 \times 5^2 \times 7 \times 11 \times 13 \times 17 \times 19 \times 23 \times 29$.<br>Divide this by $2 \times 3 \times 5 \times 7 \times 11 \times 13 \times 17 \times 19 \times 23 \times 29$, the<br>product of the primes < 30, to get $2^3 \times 3^2 \times 5 = 360$.<br><br>A) 1        B) 2        C) 12        D) 360 | 30.<br><br>D |

*The end of the contest* **A**

**Visit our Web site at http://www.mathleague.com**

# Answer Keys & Difficulty Ratings

## 2006-2007 through 2010-2011

# ANSWERS, 2006-07 7th Grade Contest

| | | | | |
|---|---|---|---|---|
| 1. C | 9. A | 17. B | 25. B | 33. D |
| 2. B | 10. C | 18. D | 26. D | 34. C |
| 3. D | 11. D | 19. A | 27. B | 35. B |
| 4. A | 12. A | 20. B | 28. B | 36. B |
| 5. A | 13. C | 21. A | 29. A | 37. A |
| 6. B | 14. B | 22. C | 30. C | 38. C |
| 7. D | 15. D | 23. D | 31. D | 39. C |
| 8. B | 16. C | 24. A | 32. D | 40. D |

# RATE YOURSELF!!!
## for the 2006-07 7th GRADE CONTEST

| Score | Rating |
|---|---|
| 38-40 | Another Einstein |
| 35-37 | Mathematical Wizard |
| 32-34 | School Champion |
| 30-31 | Grade Level Champion |
| 27-29 | Best In The Class |
| 24-26 | Excellent Student |
| 20-23 | Good Student |
| 16-19 | Average Student |
| 0-15 | Better Luck Next Time |

## ANSWERS, 2007-08 7th Grade Contest

| | | | | |
|---|---|---|---|---|
| 1. D | 9. B | 17. A | 25. C | 33. A |
| 2. C | 10. A | 18. C | 26. A | 34. C |
| 3. B | 11. D | 19. D | 27. B | 35. A |
| 4. A | 12. C | 20. A | 28. C | 36. A |
| 5. B | 13. D | 21. B | 29. B | 37. B |
| 6. D | 14. B | 22. D | 30. B | 38. C |
| 7. B | 15. C | 23. C | 31. D | 39. D |
| 8. A | 16. C | 24. D | 32. C | 40. B |

# RATE YOURSELF!!!
## for the 2007-08 7th GRADE CONTEST

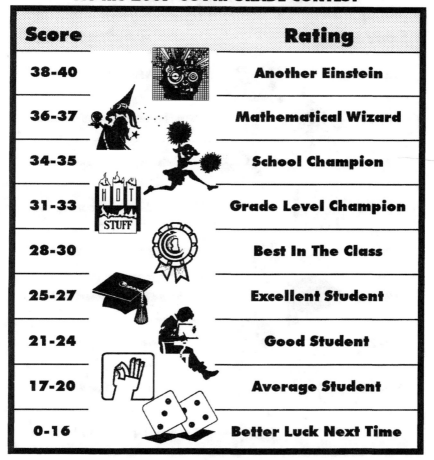

| Score | Rating |
|---|---|
| 38-40 | Another Einstein |
| 36-37 | Mathematical Wizard |
| 34-35 | School Champion |
| 31-33 | Grade Level Champion |
| 28-30 | Best In The Class |
| 25-27 | Excellent Student |
| 21-24 | Good Student |
| 17-20 | Average Student |
| 0-16 | Better Luck Next Time |

## ANSWERS, 2008-09 7th Grade Contest

| | | | | |
|---|---|---|---|---|
| 1. D | 9. A | 17. A | 25. D | 33. B |
| 2. B | 10. D | 18. B | 26. C | 34. D |
| 3. B | 11. C | 19. D | 27. A | 35. D |
| 4. C | 12. A | 20. A | 28. B | 36. A |
| 5. A | 13. D | 21. D | 29. B | 37. B |
| 6. C | 14. A | 22. D | 30. C | 38. C |
| 7. B | 15. B | 23. A | 31. B | 39. D |
| 8. C | 16. D | 24. B | 32. A | 40. C |

# RATE YOURSELF!!!
## for the 2008-09 7th GRADE CONTEST

| Score | Rating |
|---|---|
| 38-40 | Another Einstein |
| 36-37 | Mathematical Wizard |
| 33-35 | School Champion |
| 30-32 | Grade Level Champion |
| 27-29 | Best In The Class |
| 24-26 | Excellent Student |
| 20-23 | Good Student |
| 16-19 | Average Student |
| 0-15 | Better Luck Next Time |

# ANSWERS, 2009-10 7th Grade Contest

| | | | | |
|---|---|---|---|---|
| 1. B | 9. A | 17. B | 25. D | 33. D |
| 2. D | 10. D | 18. C | 26. C | 34. D |
| 3. A | 11. A | 19. C | 27. B | 35. C |
| 4. B | 12. C | 20. A | 28. D | 36. A |
| 5. C | 13. D | 21. C | 29. C | 37. B |
| 6. B | 14. A | 22. D | 30. A | 38. B |
| 7. B | 15. D | 23. A | 31. C | 39. A |
| 8. C | 16. D | 24. B | 32. B | 40. D |

# RATE YOURSELF!!!
## for the 2009-10 7th GRADE CONTEST

| Score | Rating |
|---|---|
| 39-40 | Another Einstein |
| 37-38 | Mathematical Wizard |
| 35-36 | School Champion |
| 32-34 | Grade Level Champion |
| 29-31 | Best In The Class |
| 26-28 | Excellent Student |
| 22-25 | Good Student |
| 17-21 | Average Student |
| 0-16 | Better Luck Next Time |

## ANSWERS, 2010-11 7th Grade Contest

| | | | | |
|---|---|---|---|---|
| 1. B | 8. C | 15. A | 22. A | 29. D |
| 2. D | 9. C | 16. A | 23. C | 30. B |
| 3. B | 10. B | 17. B | 24. C | 31. C |
| 4. D | 11. D | 18. C | 25. A | 32. C |
| 5. A | 12. C | 19. B | 26. A | 33. A |
| 6. A | 13. B | 20. A | 27. C | 34. B |
| 7. B | 14. D | 21. D | 28. D | 35. D |

# RATE YOURSELF!!!
## for the 2010-11 7th GRADE CONTEST

| Score | Rating |
|---|---|
| 34-35 | Another Einstein |
| 31-33 | Mathematical Wizard |
| 28-30 | School Champion |
| 25-27 | Grade Level Champion |
| 22-24 | Best In The Class |
| 20-21 | Excellent Student |
| 17-19 | Good Student |
| 14-16 | Average Student |
| 0-13 | Better Luck Next Time |

## ANSWERS, 2006-07 8th Grade Contest

| | | | | |
|---|---|---|---|---|
| 1. D | 9. D | 17. A | 25. B | 33. C |
| 2. D | 10. C | 18. C | 26. A | 34. C |
| 3. C | 11. B | 19. D | 27. D | 35. D |
| 4. B | 12. A | 20. D | 28. D | 36. A |
| 5. A | 13. D | 21. B | 29. B | 37. A |
| 6. A | 14. B | 22. C | 30. D | 38. C |
| 7. C | 15. A | 23. C | 31. D | 39. B |
| 8. A | 16. A | 24. C | 32. A | 40. B |

# RATE YOURSELF!!!
## for the 2006-07 8th GRADE CONTEST

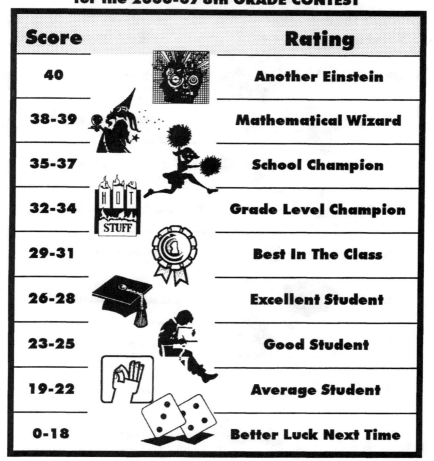

| Score | Rating |
|---|---|
| 40 | Another Einstein |
| 38-39 | Mathematical Wizard |
| 35-37 | School Champion |
| 32-34 | Grade Level Champion |
| 29-31 | Best In The Class |
| 26-28 | Excellent Student |
| 23-25 | Good Student |
| 19-22 | Average Student |
| 0-18 | Better Luck Next Time |

# ANSWERS, 2007-08 8th Grade Contest

| | | | | |
|---|---|---|---|---|
| 1. A | 9. A | 17. D | 25. D | 33. D |
| 2. D | 10. C | 18. B | 26. D | 34. A |
| 3. C | 11. B | 19. C | 27. D | 35. C |
| 4. C | 12. D | 20. A | 28. B | 36. C |
| 5. B | 13. C | 21. C | 29. A | 37. D |
| 6. B | 14. D | 22. C | 30. C | 38. B |
| 7. D | 15. C | 23. B | 31. A | 39. B |
| 8. A | 16. C | 24. A | 32. B | 40. B |

# RATE YOURSELF!!!
## for the 2007-08 8th GRADE CONTEST

| Score | Rating |
|---|---|
| 38-40 | Another Einstein |
| 36-37 | Mathematical Wizard |
| 34-35 | School Champion |
| 31-33 | Grade Level Champion |
| 28-30 | Best In The Class |
| 25-27 | Excellent Student |
| 21-24 | Good Student |
| 17-20 | Average Student |
| 0-16 | Better Luck Next Time |

# ANSWERS, 2008-09 8th Grade Contest

| | | | | |
|---|---|---|---|---|
| 1. A | 9. A | 17. A | 25. A | 33. A |
| 2. C | 10. D | 18. D | 26. D | 34. C |
| 3. B | 11. C | 19. B | 27. D | 35. A |
| 4. D | 12. C | 20. B | 28. C | 36. A |
| 5. A | 13. D | 21. D | 29. C | 37. B |
| 6. B | 14. B | 22. A | 30. D | 38. C |
| 7. C | 15. A | 23. C | 31. B | 39. A |
| 8. B | 16. B | 24. C | 32. D | 40. D |

# RATE YOURSELF!!!
## for the 2008-09 8th GRADE CONTEST

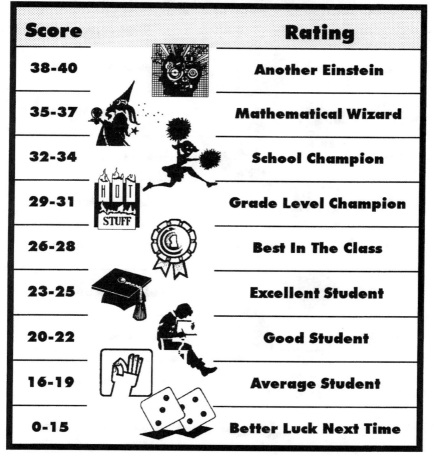

| Score | Rating |
|---|---|
| 38-40 | Another Einstein |
| 35-37 | Mathematical Wizard |
| 32-34 | School Champion |
| 29-31 | Grade Level Champion |
| 26-28 | Best In The Class |
| 23-25 | Excellent Student |
| 20-22 | Good Student |
| 16-19 | Average Student |
| 0-15 | Better Luck Next Time |

## ANSWERS, 2009-10 8th Grade Contest

| | | | | |
|---|---|---|---|---|
| 1. D | 9. A | 17. A | 25. B | 33. D |
| 2. B | 10. C | 18. D | 26. A | 34. D |
| 3. C | 11. A | 19. B | 27. A | 35. B |
| 4. B | 12. B | 20. C | 28. B | 36. A |
| 5. C | 13. A | 21. B | 29. D | 37. C |
| 6. A | 14. D | 22. D | 30. C | 38. B |
| 7. B | 15. C | 23. D | 31. B | 39. C |
| 8. D | 16. D | 24. C | 32. D | 40. A |

# RATE YOURSELF!!!
## for the 2009-10 8th GRADE CONTEST

| Score | Rating |
|---|---|
| 40 | Another Einstein |
| 38-39 | Mathematical Wizard |
| 35-37 | School Champion |
| 32-34 | Grade Level Champion |
| 29-31 | Best In The Class |
| 25-28 | Excellent Student |
| 20-24 | Good Student |
| 17-19 | Average Student |
| 0-16 | Better Luck Next Time |

## ANSWERS, 2010-11 8th Grade Contest

| | | | | |
|---|---|---|---|---|
| 1. C | 8. A | 15. B | 22. C | 29. D |
| 2. A | 9. B | 16. A | 23. C | 30. B |
| 3. C | 10. D | 17. B | 24. D | 31. D |
| 4. A | 11. B | 18. B | 25. C | 32. A |
| 5. D | 12. C | 19. A | 26. A | 33. D |
| 6. D | 13. D | 20. B | 27. B | 34. C |
| 7. C | 14. A | 21. B | 28. B | 35. D |

# RATE YOURSELF!!!
## for the 2010-11 8th GRADE CONTEST

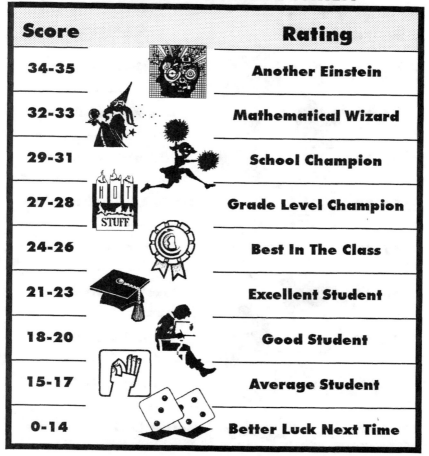

| Score | Rating |
|---|---|
| 34-35 | Another Einstein |
| 32-33 | Mathematical Wizard |
| 29-31 | School Champion |
| 27-28 | Grade Level Champion |
| 24-26 | Best In The Class |
| 21-23 | Excellent Student |
| 18-20 | Good Student |
| 15-17 | Average Student |
| 0-14 | Better Luck Next Time |

## ANSWERS, 2006-07 Algebra Course 1 Contest

| | | | | |
|---|---|---|---|---|
| 1. D | 7. A | 13. B | 19. C | 25. A |
| 2. D | 8. D | 14. A | 20. A | 26. B |
| 3. C | 9. B | 15. C | 21. C | 27. A |
| 4. A | 10. D | 16. D | 22. D | 28. C |
| 5. B | 11. C | 17. C | 23. C | 29. B |
| 6. B | 12. D | 18. B | 24. D | 30. B |

# RATE YOURSELF!!!
## for the 2006-07 ALGEBRA COURSE 1 CONTEST

| Score | Rating |
|---|---|
| 28-30 | Another Einstein |
| 24-27 | Mathematical Wizard |
| 21-23 | School Champion |
| 19-20 | Grade Level Champion |
| 17-18 | Best In The Class |
| 14-16 | Excellent Student |
| 12-13 | Good Student |
| 9-11 | Average Student |
| 0-8 | Better Luck Next Time |

# ANSWERS, 2007-08 Algebra Course 1 Contest

| | | | | |
|---|---|---|---|---|
| 1. B | 7. D | 13. A | 19. C | 25. B |
| 2. C | 8. C | 14. A | 20. D | 26. C |
| 3. C | 9. B | 15. D | 21. C | 27. D |
| 4. A | 10. D | 16. A | 22. C | 28. D |
| 5. D | 11. A | 17. D | 23. A | 29. C |
| 6. B | 12. B | 18. B | 24. B | 30. A |

# RATE YOURSELF!!!
## for the 2007-08 ALGEBRA COURSE 1 CONTEST

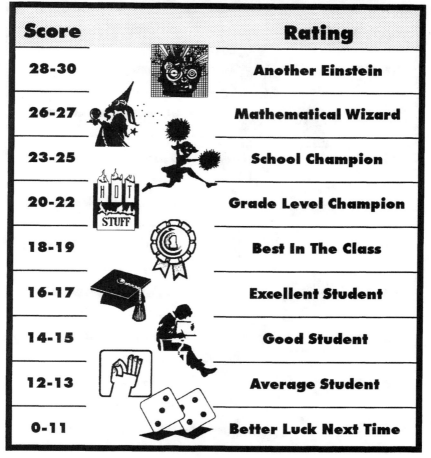

| Score | Rating |
|---|---|
| 28-30 | Another Einstein |
| 26-27 | Mathematical Wizard |
| 23-25 | School Champion |
| 20-22 | Grade Level Champion |
| 18-19 | Best In The Class |
| 16-17 | Excellent Student |
| 14-15 | Good Student |
| 12-13 | Average Student |
| 0-11 | Better Luck Next Time |

# ANSWERS, 2008-09 Algebra Course 1 Contest

| | | | | |
|---|---|---|---|---|
| 1. D | 7. D | 13. C | 19. A | 25. C |
| 2. C | 8. A | 14. A | 20. A | 26. B |
| 3. B | 9. B | 15. B | 21. D | 27. C |
| 4. A | 10. C | 16. D | 22. D | 28. B |
| 5. B | 11. D | 17. D | 23. A | 29. D |
| 6. C | 12. C | 18. B | 24. A | 30. C |

# RATE YOURSELF!!!
## for the 2008-09 ALGEBRA COURSE 1 CONTEST

| Score | Rating |
|---|---|
| 28-30 | Another Einstein |
| 25-27 | Mathematical Wizard |
| 21-24 | School Champion |
| 19-20 | Grade Level Champion |
| 17-18 | Best In The Class |
| 15-16 | Excellent Student |
| 13-14 | Good Student |
| 10-12 | Average Student |
| 0-9 | Better Luck Next Time |

# ANSWERS, 2009-10 Algebra Course 1 Contest

| | | | | |
|---|---|---|---|---|
| 1. D | 7. B | 13. A | 19. B | 25. C |
| 2. A | 8. D | 14. C | 20. B | 26. A |
| 3. A | 9. B | 15. D | 21. A | 27. C |
| 4. C | 10. A | 16. D | 22. D | 28. B |
| 5. A | 11. D | 17. A | 23. C | 29. A |
| 6. B | 12. B | 18. C | 24. D | 30. B |

# RATE YOURSELF!!!
## for the 2009-10 ALGEBRA COURSE 1 CONTEST

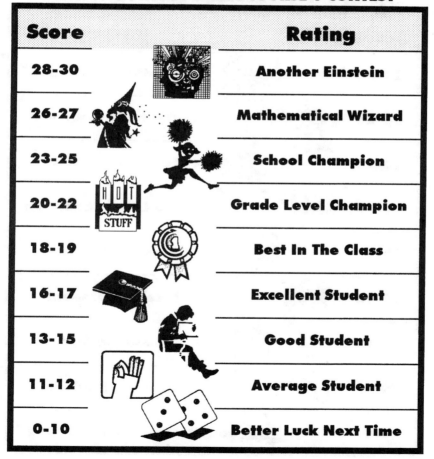

| Score | Rating |
|---|---|
| 28-30 | Another Einstein |
| 26-27 | Mathematical Wizard |
| 23-25 | School Champion |
| 20-22 | Grade Level Champion |
| 18-19 | Best In The Class |
| 16-17 | Excellent Student |
| 13-15 | Good Student |
| 11-12 | Average Student |
| 0-10 | Better Luck Next Time |

# ANSWERS, 2010-11 Algebra Course 1 Contest

| | | | | |
|---|---|---|---|---|
| 1. D | 7. B | 13. A | 19. B | 25. C |
| 2. C | 8. A | 14. C | 20. D | 26. B |
| 3. C | 9. D | 15. A | 21. A | 27. A |
| 4. A | 10. D | 16. C | 22. D | 28. A |
| 5. B | 11. B | 17. B | 23. B | 29. A |
| 6. C | 12. C | 18. D | 24. D | 30. D |

# RATE YOURSELF!!!
## for the 2010-11 ALGEBRA COURSE 1 CONTEST

| Score | Rating |
|---|---|
| 27-30 | Another Einstein |
| 24-26 | Mathematical Wizard |
| 20-23 | School Champion |
| 18-19 | Grade Level Champion |
| 16-17 | Best In The Class |
| 14-15 | Excellent Student |
| 11-13 | Good Student |
| 9-10 | Average Student |
| 0-8 | Better Luck Next Time |

# Math League Contest Books
## 4th Grade Through High School Levels

Written by Steven Conrad & Daniel Flegler, recipients of Ronald Reagan's 1985 Presidential Awards for Excellence in Mathematics Teaching
- *Easy-to-use format designed for a 30-minute period*
- *Problems ranging from straightforward to challenging*

Order books at www.mathleague.com (or use the form below)

Name: _____

Address: _____

City: _____ State: _____ Zip: _____

| Available Titles | # of Copies | Cost |
|---|---|---|
| **Math Contests—Grades 4, 5, 6** | ($12.95 each) | |
| Volume 1: 1979-80 through 1985-86 | _____ | _____ |
| Volume 2: 1986-87 through 1990-91 | _____ | _____ |
| Volume 3: 1991-92 through 1995-96 | _____ | _____ |
| Volume 4: 1996-97 through 2000-01 | _____ | _____ |
| Volume 5: 2001-02 through 2005-06 | _____ | _____ |
| Volume 6: 2006-07 through 2010-11 | _____ | _____ |
| **Math Contests—Grades 7 & 8 ‡** | ‡(Vols. 3,4,5,6 include Alg. Course I) | |
| Volume 1: 1977-78 through 1981-82 | _____ | _____ |
| Volume 2: 1982-83 through 1990-91 | _____ | _____ |
| Volume 3: 1991-92 through 1995-96 | _____ | _____ |
| Volume 4: 1996-97 through 2000-01 | _____ | _____ |
| Volume 5: 2001-02 through 2005-06 | _____ | _____ |
| Volume 6: 2006-07 through 2010-11 | _____ | _____ |
| **Math Contests—High School** | | |
| Volume 1: 1977-78 through 1981-82 | _____ | _____ |
| Volume 2: 1982-83 through 1990-91 | _____ | _____ |
| Volume 3: 1991-92 through 1995-96 | _____ | _____ |
| Volume 4: 1996-97 through 2000-01 | _____ | _____ |
| Volume 5: 2001-02 through 2005-06 | _____ | _____ |
| Volume 6: 2006-07 through 2010-11 | _____ | _____ |
| ***Shipping and Handling*** | $3 ($5 Canadian) | |
| *Please allow 2-4 weeks for delivery* | Total: $_____ | |

☐ Check or Purchase Order Enclosed; **or**

☐ Visa / MC/Amex # _____ Expires _____

☐ Security Code _____ Signature _____

Mail your order with payment to:
**Math League Press. PO Box 17, Tenafly, New Jersey USA 07670–0017**
**or order on the Web at www.mathleague.com**

Phone: (201) 568-6328 • Fax: (201) 816-0125